DOCTOR WHO

The ShaKespeare

NOTEBOOKS

James Goss has written the books *Doctor Who: Dead of Winter* and *Summer Falls*, as well as several *Torchwood* books and radio plays. His favourite play is *Pericles*, oddly.

Jonathan Morris is one of the most prolific authors of *Doctor Who* novels, audio plays and comic strips and a regular contributor to *Doctor Who Magazine*. He first read the complete works of Shakespeare as a precocious 15-year-old but 25 years later still struggles with iambic pentameter.

Julian Richards is an English and Theatre Studies student at the University of Warwick. He is a Third Dan Karate Instructor, keen amateur writer and lifelong fan of both Shakespeare and *Doctor Who*, making this almost his ideal book (lacking only Karate).

Justin Richards has written for stage and screen, audio, children's novels, the science fiction series *The Never War*, and all sorts of other things. In his spare time he acts as Creative Consultant to BBC Books for all their *Doctor Who* titles – including this one. He has a degree in English and Theatre Studies, and once kidnapped someone by accident.

Matthew Sweet is a writer and broadcaster with a doctorate in Wilkie Collins. He presents *Free Thinking* and *Sound of Cinema* on BBC Radio 3 and *The Philosopher's Arms* on BBC Radio 4. His books and TV documentaries include *The West End Front*, *Shepperton Babylon*, *The Rules of Film Noir* and *Me, You and Doctor Who*. A million years ago he played Iachimo to Eve Best's Imogen, but Hollywood never called.

BBC

DOCTOR WHO

The Shakespeare
NOTEBOOKS

HARPER
DESIGN
An Imprint of HarperCollins Publishers

Text by James Goss, Jonathan Morris, Julian Richards, Justin Richards,
William Shakespeare and Matthew Sweet, with additional thanks to
Becca Dunn, Jenni Street and Helen Cornes © 2014
Illustrations by Mike Collins © Woodlands Books Ltd, 2014

Doctor Who is a BBC Wales production for BBC One.
Executive producers: Steven Moffat and Brian Minchin

First published in 2014 by:
Harper Design
An Imprint of HarperCollins*Publishers*
195 Broadway
New York, NY 10007
Tel: (212) 207-7000
Fax: (212) 207-7654

Ditributed throughout the world by:
HarperCollins*Publishers*
195 Broadway
New York, NY 10007

Library of Congress Number: 2014939083
ISBN 978-0-06-234442-7

Editorial director: Albert DePetrillo
Editorial managers: Lizzy Gaisford and Joe Cottington
Series consultant: Justin Richards
Project editor: Steve Tribe
Illustrations: Mike Collins
Design: Seagull Design
Cover design: Two Associates © Woodlands Books Ltd 2014
Production: Alex Goddard

Printed and bound in the United States.

First printing, 2014.

CONTENTS

Doctor Who? That is the question.

PREFACE TO
THE FIRST EDITION

Considering he is acknowledged as the world's greatest playwright, surprisingly little is actually known about the life and times of William Shakespeare. We know the man almost entirely through his work, and the occasional accounts of others. It is possible that the most famous portrait of the man is not actually of Shakespeare at all. Despite the evidence of his genius, he remains a mystery.

So it was with excitement, trepidation, and also a degree of scepticism within the academic world when the so-called Shakespeare Notebooks were discovered recently. Could they be genuine? And if so, what did they *mean* – not just for our understanding of the man's life but also of his work? On the face of it, the Notebooks are a record kept by Shakespeare himself over several years. Not so much a journal as a scrapbook containing early drafts of key scenes and moments from his plays, as well as other observations, and previously unknown material, including several sonnets.

In literary terms, the Notebooks are priceless. But for many it is the single page of explanation at the front of the Notebooks that is most intriguing and enigmatic. Is it indeed by Shakespeare himself? If so, to what events does it refer? We reproduce the text of that preface here for the first time in its entirety. Note that the 'day book' (or diary) to which Shakespeare refers at the start of the text has never been found, and so the references he describes remain an enigma.

Following the strange and unsettling events of the first, and of necessity the only, staging of my play *Love's Labour's Won*, I, William Shakespeare, have recorded the incident in my day book. As I put down upon the page what happened, and recorded my thoughts of the mysterious stranger known as 'the Doctor', it occurred to me that this was not the first time we had met.

'Though his face and form seemed unfamiliar, freed from the baleful influence of the Carrionites, I have come to realise that several other mysterious strangers who have influenced my life and my work may all in point of fact be this one and the same person.

And so, inspired by this revelation, I have traversed the history of my notes and journals. From these and other divers places have I compiled this book of scraps. A volume wherein I do draw together every incident and encounter that may perchance have involved or been influenced by the Doctor. It has been an enlightenment, and I have found the Doctor to have appeared not only in my life, but in my writings too. Can I have forgot so completely whereof my inspirations came, and thought them but the dew of imagination, the sweat upon the brow of diligence and labour?

It is as if the Doctor has somehow traversed my life in retrospect, removing any references and allusions to himself and to the strange world of wonders and magick that is his habitation. And now, save for the sundry items I do gather here, these recollections do remain only in my most private thoughts and the fading tablet of my memory...

Shakespeare seems to have become somewhat obsessed with the enigmatic 'doctor' to whom he refers. Within the Notebooks this figure seems to appear in various guises – as magician, physician, academic, colleague and friend, and in some extracts merely as 'the Man' or 'He'. But who he might have been – assuming he even existed – is never fully explained. Perhaps the Notebooks themselves formed the basis for a possible epic work in which Shakespeare planned to present the adventures of this 'doctor'.

But whatever Shakespeare's plans and aspirations, on his death, the Notebooks – the Bard's final words on the subject of the enigmatic 'doctor' – were also lost. They remained hidden for four hundred years.

Until now.

Finally, we have obtained the rights to the original Notebooks put together by William Shakespeare in late 1599 and kept updated until his death – the scrapbook in which he documented the impact of the 'doctor' on his life and work. It is a strange collection of writings, snippets, and musings.

Here for the first time we can see the original notes for *Hamlet*, including a very different appearance by the ghost. We can ponder on early versions of great lines from the Bard and wonder why they were changed. We can see how the faeries of *A Midsummer Night's Dream* were originally imagined, and how the stage directions were later adjusted to remove references to a mysterious blue box and 'the strangest sound akin to wheezing and groaning that ever did assail the ear'.

This special edition of *The Shakespeare Notebooks* presents material not only from the Notebooks themselves, but also from a variety of other contemporary and more modern sources. For the first time, we bring together other material that would seem to have a direct bearing on the mysterious 'doctor'. The editors have organised the book into the following sections:

The Shakespeare Notebooks

This section contains the material gathered together by Shakespeare himself in the Notebooks. Most of it he had written himself, although some may be drawn from other sources. Here you will find early drafts of his work which predate the versions we are so familiar with today. This section also includes material pertaining to Shakespeare's work in performance, or in what he considered to be 'final form', together with some contemporary Elizabethan and Jacobean material relevant to the plays.

The Timeless Shakespeare

This section presents material generally taken from later years, right up to the present day, that would seem to elaborate on Shakespeare's apparent contention that the strange 'doctor' was somehow unbounded by the limitations of time. It includes post-Shakespearean material such as reviews of performances, critical reception, and academic analysis of the work.

Interspersed with the main extracts, we have included some of the single-line 'drafts' that occur in the margins throughout the Notebooks. There are, as you can imagine, many of these, so we have tended towards the lines that are well known in their final form.

Note that, so far as possible, the original inconsistent layout and presentation of the source material has been preserved. That said, each of the main extracts is prefaced with explanatory text. But in the final analysis is it up to you, the reader, to determine whether you believe the Shakespeare Notebooks are indeed genuine, or an elaborate hoax.

I am but mad north-north-west.
When the time wind's southerly,
I know a Dalek from a Cyberman.

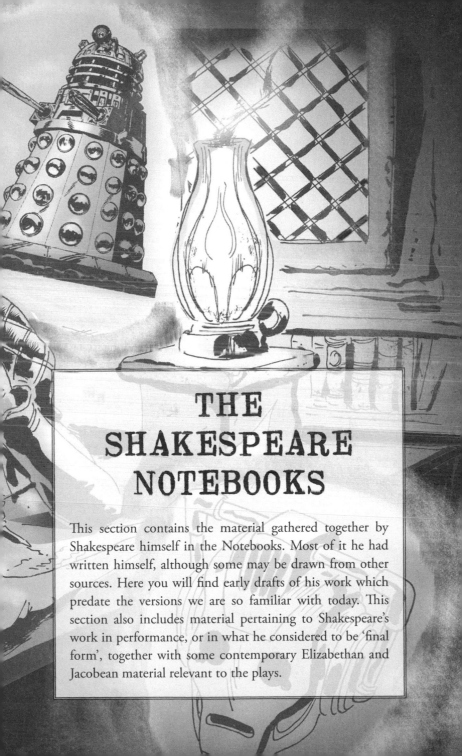

THE SHAKESPEARE NOTEBOOKS

This section contains the material gathered together by Shakespeare himself in the Notebooks. Most of it he had written himself, although some may be drawn from other sources. Here you will find early drafts of his work which predate the versions we are so familiar with today. This section also includes material pertaining to Shakespeare's work in performance, or in what he considered to be 'final form', together with some contemporary Elizabethan and Jacobean material relevant to the plays.

NOTES ON A PLAY

These handwritten notes on a single sheet within the Notebooks seem to be Shakespeare's first thoughts on the early scenes of Hamlet. Note that the ghost of Hamlet's father was to take on a rather different aspect originally.

Elsinore – is that in Denmark? 'Hamlet, Prince of Finland' doesn't have quite the same gravitas.

King – dead, killed by brother who marries mother and takes the throne. Laws of succession are quite odd in Denmark (not sure about Finland).

Hamlet – son of dead king (hence 'Prince of Denmark') discovers his uncle murdered his father and married his mother to steal the throne. Probably drives him mad. Well, it would, wouldn't it?!

How's he find out? Possibilities:

- Hidden papers – never very satisfactory
- His mother confesses – except she wouldn't know, would she?
- His uncle confesses – why? That's asking for trouble!
- He guesses – hmmm.
- A magician tells him – ah, most plausible. Such things are common in the theatre.

The magician appears in a strange blue box announced by all the sounds of hell itself. Yes, that could work. Magician wears exotic apparel – a knotted bow of ribbon at his neck, for instance. Perhaps like this:

A blue box doth appear and Magician enters from box

HAMLET
>Angels and ministers of grace defend us!
>Be thou a spirit of health or goblin damn'd,
>Bring with thee airs from heaven or blasts from hell,
>Be thy intents wicked or charitable,
>Thou com'st in such a questionable shape
>That I will speak to thee.

MAGICIAN
>What country, sir, is this?

HAMLET
>Why, 'tis Denmark, sire. Or mayhap Finland. What manner of man art thou?

MAGICIAN
>I am Magician. Do you not know me, for we have met before, Lord Hamlet.

HAMLET
>Whither wilt thou lead me? Speak! I'll go no further.

MAGICIAN
>Mark me.

HAMLET
>I will.

MAGICIAN
>My hour is almost come,
>When I to Vortex and tormenting Time
>Must render up myself.

HAMLET
>Alas, poor friend!

MAGICIAN
Pity me not, but lend thy serious hearing
To what I shall unfold.

HAMLET
Speak. I am bound to hear.

MAGICIAN
So art thou to revenge, when thou shalt hear.

HAMLET
What?

MAGICIAN
I am a Lord of Time,
Doom'd for all eternity to walk,
And for the day confin'd to my blue box,
Till the foul crimes done in the War of Time
Are burnt and purg'd away. But that I am forbid
To tell the secrets of my home world,
I could a tale unfold whose lightest word
Would harrow up thy soul, freeze thy young blood,
Make thy two eyes, like stars, start from their spheres,
Thy knotted and combined bow tie to part,
And each particular hair to stand on end
Like quills upon Koquillion.
But this eternal blazon must not be
To ears of flesh and blood. List, list, O, list!
If thou didst ever thy dear father love –

HAMLET
O God!

MAGICIAN
Revenge his foul and most unnatural murther.

Needs a bit of work still, obviously. Also – note to self – check spelling of 'murther'.

EXITS AND ENTRANCES

An early draft of one of Shakespeare's most famous speeches, this draft from As You Like It *appears without explanation within the Notebooks. It was probably revised as many of the allusions seem rather obscure.*

The cosmos is a stage,
And all, from thou to Fenric, merely players:
We have our exits and our entrances;
But Time Lords are required to play the greater part,
My acts being seven ages. At first the ancient,
Pondering caveman's eggshell skull.
And then the stovepiped clown in pixie boots,
Tripping twixt pools of noisome mercury
Heedlessly on Vulcan. And then the gent,
Sighing at circuits, with a woeful ballad
Made over lifeless switches. Then the scarf
Inside which lives a man of Left Bank tastes:
A wand'rer in the fourth dimension,
Given to sudden thunder. Then the youth,
A sporting chap with linseed on his cuffs
Who keeps to time much like a broken clock
(Which has authority but twice a day.)
And so he plays his part. The sixth age slips
Into some mustard-coloured pantaloons.
A man who shouts upon the mountain top
Throws quotes at slugs and wrings the necks of friends,
And woe betide the guard who fights with him
Beside a tank of *aqua regia*,
With quip as epitaph. Last scene of all,
That ends this strophe of mine own history,
Seems stranger still as I survey this board,
And play for thee, and me, and everything.

THE TRUE TRAGEDIE OF MACBETH

It is well known that Macbeth *is the shortest of Shakespeare's tragedies, leading many critics to speculate that the text first printed in the Folio was a heavily edited version of what was originally a play of similar length to* Othello *and* King Lear. *Certainly there is evidence that the text contains interpolations drawn from the work of Thomas Middleton and that some scenes may have been heavily cut (see Wilson, 1947).*

The Shakespeare Notebooks do not contain a complete version of the play, but give us some tantalising glimpses of what it may have contained, as it features alternative versions of several scenes along with some omitted material. It seems that Shakespeare originally intended for the play (then entitled The True Tragedie of Macbeth) *to contain an additional subplot and three extra characters. It seems likely they were omitted from the finished play for reasons of staging, of dramatic unity, and of plausibility. What is interesting, though, is that Shakespeare seems to have been working from a source for the play in addition to Holinshed's Chronicles...*

ACT I, SCENE III – A HEATH

Thick fog. Trumpet, wheezing, groaning.
DOCTOR, JAMIE and ZOE appear.

JAMIE

    Och, Doctor. What've you gone and done now?

DOCTOR

    I'm not sure, Jamie. Some small malfunction
    Of Tardis circuits I expect.

JAMIE

'Some small'? One minute we're in the control room, the
next we're on some blasted heath.

ZOE

Where do you think we are then, Doctor, pray?

DOCTOR

Afraid I don't know, Zoe. 'Tis too foul
To navigate by stars. On Earth, I guess.

JAMIE

Aye, it always is!

ZOE

And 'tis the mid of night. And raineth hard.

Thunder

JAMIE

We'll freeze to death out here. We should find shelter.

DOCTOR

I do not think that would be very wise.

JAMIE

Why not?

DOCTOR

For at some point the fault will mend itself
And we to Tardis shall be swift return'd.

ZOE

So long as we stand fast upon this spot?

DOCTOR

Indeed, my dear.

JAMIE

You're saying that if we do move, we might not get back?

DOCTOR

Besides, who knows what hazards this fog veils?

ZOE

Yes. I can't see more than an arm's length hence –

Zoe trips.

JAMIE

Zoe? Are you all right?

ZOE

I think so. I just tripp'd over this bush.

JAMIE

Hey, Doctor! It's heather.

DOCTOR

So?

JAMIE

So we might be in Scotland!

ZOE

Forsooth, 'tis wet enough it must be said. *[Sound of approaching soldiers]*

DOCTOR

Hush! Hark! I hear approaches through the dark
Quick, Zoe, Jamie, get down out of sight!

JAMIE

Och no, in the mud?

They fall in the mud.

ZOE

Too late, by them I think we have been glimps'd!

Enter MACBETH and BANQUO.

MACBETH

So foul and fair a day I have not seen.

BANQUO

How far is't to Forres? *[Sees the DOCTOR, JAMIE and ZOE]*
What are these,
So muddied and so wild in their attire,
That look not like th'inhabitants o'th'earth,
And yet are on't?

JAMIE

Hey, are you talking about us?

BANQUO

My lord Macbeth, they speak!

*The DOCTOR, JAMIE and ZOE get up,
covered in mud.*

ZOE

Macbeth? Did I just hear you say Macbeth?

JAMIE

Hey, I've heard of him! He was the Thane of Glamis back in
olden times. And the Thane of Cawdor.

DOCTOR

Hush! Jamie, hush!

ZOE

There was a play about him I recall
Did not he become King of Scotland too?

MACBETH

[amazed] You say I shall be King hereafter?

BANQUO

Good Sir, why do you start, and seem to fear
Things that do sound so fair?
[to the DOCTOR]
If you can look into the seeds of time,
And say which grain will grow, and which will not,
Speak then to me, Banquo, who neither fears,
Your favours nor your hate.

JAMIE

You're Banquo! I've heard of you too!

DOCTOR

Please, Jamie, keep quiet, I beg of you!

JAMIE

[to BANQUO] You will beget kings, though you won't be
one.

MACBETH

Stay, you imperfect speakers, tell me more.
By Sinel's death I know I am Thane of Glamis;
But how of Cawdor? The Thane of Cawdor lives,
A prosperous gentleman; and to be King
Stands not within the prospect of belief,
No more than to be Cawdor. Say from whence
You owe this strange intelligence? Or why
Upon this blasted heath you stop our way
With such prophetic greeting? – Speak, I charge you.

JAMIE

We know because we're –

DOCTOR clamps his hand over JAMIE's mouth.
They struggle. Trumpet, wheezing, groaning.

ZOE

Oh Doctor, we are disappearing fast!

DOCTOR

Hold tight! For we arc being taken back!

The DOCTOR, JAMIE and ZOE vanish.

BANQUO

Whither are they vanish'd?

MACBETH

Into the air; and what seem'd corporal,
Melted as breath into the wind. Would they had stay'd!
Your children shall bc kings.

BANQUO

You shall be King.

MACBETH

And Thane of Cawdor too…

ACT II, SCENE III – OUTSIDE THE CASTLE

Heavy rain. Trumpet, wheezing, groaning. TARDIS
appears. DOCTOR, JAMIE and ZOE emerge having
cleaned up, wearing heavy, hooded coats.

JAMIE

At least we've landed properly this time.

ZOE

We've mov'd?

DOCTOR

Yes, Zoe. That's Macbeth's castle!

JAMIE

Ah, yon fellow we met.

DOCTOR

Indeed, the one you told his destiny.
I wish you had not spoke of that, alas!

JAMIE

Och, why? What harm could it do?

DOCTOR

What harm? Now he will want to slay the King
And take his sov'reign throne!

ZOE

Like in the play!

JAMIE

What play?

DOCTOR

The play, Macbeth.

The DOCTOR takes a book from his pocket.

DOCTOR

Now let me look. Ah, yes, now here it is.
Macbeth is made the Thane of Cawdor first,
Which makes him heed the prophecy and thence,
With nudging from his wife, he kills the King.
Oh dear.

JAMIE

Oh dear?

DOCTOR

It's all our fault! *[DOCTOR slaps JAMIE with the book]*
Your fault!
We must this mend before it is too late!

DOCTOR puts book away and knocks on castle gates.

DOCTOR

Hello! Pray let us in! Hello! Hello!

ZOE

But Doctor, if Shakespeare didst write this tale,
It must therefore be true, and come to pass?

DOCTOR

My dear, you should not heed what's writ by Will
He chang'd things round to serve his own intents,
The play's the thing, it's what all poets do,
They take poetic license with the truth.

JAMIE

But I know Macbeth became King, it's in the history books.

DOCTOR

But not by murder necessarily!

DOCTOR knocks on castle gates.

DOCTOR

Come on! Pray open up! We're soaking wet!

JAMIE

Aye, it's raining cats and dogs out here.

ZOE

And raging wind that bites and roars so loud!

JAMIE
> Aye, sounds like the devil himself. And the ground's
> shaking –

ZOE
> Look to't! One of the chimneys is disturb'd!

It crashes down beside them. DOCTOR knocks on castle gates.

DOCTOR
> Come on! You're in grave danger! Open up!

Castle gates open. PORTER appears.

PORTER
> Knock knock, who's there, i'th'name of Belzebub?

DOCTOR
> The Doctor!

PORTER
> Doctor? Who's that? Doctor who?

DOCTOR
> That matters not, what matters is that if
> Thou dost not let us in there will be deeds
> Most foul perform'd to-night within these walls!

PORTER lets them into the castle.

PORTER
> Anon, anon!
> I pray you, remember the Porter!

*The DOCTOR, JAMIE and ZOE enter a court
within the castle.*

DOCTOR
Now, we must find the King and warn him what
Macbeth has plann'd! There is no time to lose!

PORTER
This place is too cold for Hell. I'll devil-porter it no
Further.

*The PORTER goes to leave but is stopped as MACDUFF
and LENOX enter. The DOCTOR, JAMIE and ZOE
duck out of sight.*

MACDUFF
[to the PORTER] Is thy master stirring?

Enter MACBETH.

The knocking has awak'd him; here he comes.

LENOX
Good morrow, noble Sir!

MACBETH
Good morrow, both!

In the background, the DOCTOR checks the book.

MACDUFF
Is the King stirring, worthy Thane?

MACBETH
Not yet.

The Doctor reads something that alarms him.

MACDUFF

> He did command me to call timely on him:
> I have almost slipp'd the hour.

MACDUFF exits, going into the King's chambers.

ZOE

> *[whispers]* Doctor, what's the matter?

DOCTOR

> *[whispers]* According to the play, we are too late
> Already vile Macbeth has slain the King!

Re-enter MACDUFF.

MACDUFF

> O horror! Horror! Horror!
> Tongue nor heart cannot conceive, nor name thee!

MACBETH, LENOX

> What's the matter?

MACDUFF

> Most sacrilegious Murther hath been wrought
> See, and then speak yourselves. –

MACBETH and LENOX go to the King's chamber.

> Awake! Awake! –
> Ring the alarum-bell. – Murther, and treason!
> Banquo, and Donalbain! Malcolm, awake!
> Shake off this downy sleep, death's counterfeit,
> As from your graves rise up, and walk like sprites,
> To countenance this horror! *[Bell rings]*

Enter LADY MACBETH.

LADY MACBETH
>
> What the business,
> That such a hideous trumpet calls to parley
> The sleepers of the house? Speak, speak!

ZOE
>
> *[whispers]* Doctor! She must be Lady Macbeth! Look!

JAMIE
>
> The one who put her husband up to it? Look at her, acting
> like she doesn't know what's going on!

DOCTOR
>
> Quick, Jamie, Zoe, we should slip away!

JAMIE
>
> But Doctor –

ZOE
>
> The Doctor's right, we don't want to be caught
> They'll think we were the ones who did the deed!

> *The DOCTOR, JAMIE and ZOE slip away as*
> *MACBETH and LENOX return from the*
> *King's chamber.*

MACBETH
>
> Had I but died an hour before this chance,
> I had liv'd a blessed time; for, from this instant,
> There's nothing serious in mortality;
> All is but toys: renown, and grave, is dead!

> *Enter MALCOLM and DONALBAIN.*

DONALBAIN
>
> What is amiss?

MACDUFF
>Your royal father's murther'd!

ACT II, SCENE V – A PASSAGE IN THE CASTLE

The DOCTOR, JAMIE and ZOE hide.

DOCTOR
>All right, I think we're safe.

ZOE
>Until they find
>Us and accuse us of murther – murder!

DOCTOR
>Be not afear'd. *[the Doctor checks his book]* According to
>the play
>Two guards are blam'd and art by Macbeth slain.

JAMIE
>I thought you said the play wasn't right?

ZOE
>But nobody is going to believe
>That it was not by Macbeth's hand achiev'd.

JAMIE
>Aye, he had guilt written all over his face.

DOCTOR
>You may be right. But now he'll reign as King
>Of Scotland, as you told him would ensue.

JAMIE
>You can't blame me for that!

DOCTOR

What matters is t'avert more lives be lost.
You told Banquo that would he Kings beget
Which means he is a threat unto Macbeth.

ZOE

Macbeth will have to put him out the way?

DOCTOR

Unless we can prevent the misdeed, yes.

JAMIE

But why do we have to interfere? Can't we just let history get
on with it?

DOCTOR

Because we set this train upon its track
It's down to us to see no ill proceeds.

ZOE

What must we do I keenly bid you tell.

Enter a SERVANT.

SERVANT

Are you the travellers lately arriv'd?

JAMIE

Might be, why do you want to know?

SERVANT

The King desires your presence.

ZOE

The King? You mean?

SERVANT

> King Macbeth. He requires your services, that is why you
> were summon'd. If you will come with me –

DOCTOR

> Ah, Zoe, wait here 'til we are return'd.

ZOE

> But –

JAMIE

> The Doctor's right, you're safer staying put.

ZOE

> Oh, very well.

DOCTOR

> We won't be long, I pledge
> I wonder what the King should want us for…

> *The SERVANT leads the DOCTOR and
> JAMIE into the castle.*

ACT III, SCENE I – THE CASTLE THRONE ROOM

> *MACBETH sits on the throne. The SERVANT enters
> with the DOCTOR and JAMIE and leaves. MACBETH
> regards the DOCTOR and JAMIE suspiciously. They both
> have their faces concealed by their hoods.*

MACBETH

> *[suspicious]* Sirs, have I met with you afore this day?

DOCTOR

> We have not had the honour, Majesty.

MACBETH

Now, if you have a station in the file,
Not i'th'worst rank of manhood, say't;
And I will put that business in your bosoms,
Whose execution takes your enemy off,
Grapples you to the heart and love of us,
Who wear our health but sickly in his life,
Which in his death were perfect?

JAMIE

[whisper] Doctor, what's he going on about?

DOCTOR

[whisper] He's asking if we are unscrupulous
Villains prepar'd to kill to please their king.

JAMIE

What?

DOCTOR

He thinks we are two murderers for hire.
Attend and hush! *[clears throat, to MACBETH in a gruff
accent of Glasgow]* I am one such, my Liege;
The many slings and arrows of the world
Hath so incens'd, that I am reckless what
I do, to spite the world.

JAMIE

[gruff Glaswegian] Aye, and what he just said goes for me too.

MACBETH

You care not for your own lives? Then you should
Know Banquo is mine foe. And thence it is
That I to your assistance do request,
Masking the business from the common-eye,
For sundry weighty reasons.

DOCTOR
We shall, my Lord,
Perform what you command us.

JAMIE
So do you want us to do Banquo in, then?

MACBETH
[nods and smiles] Your spirits shine through you. Within this
hour, at most,
I will advise you where to plant yourselves,
Acquaint you with the perfect spy o'th'time,
The moment on't; for't must be done to-night.

JAMIE
Tonight!

MACBETH
It is concluded: Banquo, thy soul's flight,
If it find Heaven, must find it out to-night!

Exit MACBETH.

JAMIE
Doctor, what's going on? We've just agreed to kill Banquo!

DOCTOR
I know. Rather the King ask us than two
Real murderers, that would you not agree?

JAMIE
What?

DOCTOR
By hiring us to cut short Banquo's life,
He's granted us a means of thwarting strife!

Exeunt the DOCTOR and JAMIE.

ACT III, SCENE III – THE ROAD LEADING FROM THE CASTLE

Enter the DOCTOR, JAMIE and ZOE.

ZOE

Macbeth ask'd you to Banquo slay? To this
You did complot?

DOCTOR

Yes, to secure his 'scape.

JAMIE

It's getting dark.

ZOE

The west yet glimmers with the some streaks of day.
Hark! I hear men walk this way. A light!

DOCTOR

'Tis he!

*Enter BANQUO and his son FLEANCE,
carrying torches.*

BANQUO

It will be rain to-night.

DOCTOR

The time to act is nigh. Stand to, with me!

The DOCTOR, JAMIE and ZOE approach them.

BANQUO

Who is this dar'st waylay us 'pon the road?

DOCTOR

Macbeth dispatch'd us here to murder you.

BANQUO
[draws sword] What?

ZOE
Prithee, be not alarm'd! We are unarm'd!

BANQUO
You have no swords?

DOCTOR
We wield no thought of harm.
We've come to deliver you from death's reach.

BANQUO
You are still true to me, then? Not Macbeth?

DOCTOR
No loyalty to him we owe; he is
A regicidal maniac. We will
Attest we kill'd you. Fleance must away
To Wales, a fine girl meet, and Kings beget.

FLEANCE
I flee to Wales? But what of my father?

DOCTOR
He'll follow you anon. I need his help.

BANQUO
With what?

DOCTOR
The King shall not be satisfied
Thou art slain without proof. Thus I intend
That proof to give. With this subtle device.

The DOCTOR hands BANQUO a necklace.

BANQUO
An amulet?

DOCTOR
Something I pick'd up on
My travels. It's a perception filter.
It can make you invisible or make
It that but one soul wilt of you perceive
You will be like a ghost at Macbeth's feast…

ACT III, SCENE IV – THE CASTLE BANQUETING HALL

*MACBETH, LADY MACBETH, ROSSE, LENOX,
Lords and Attendants are enjoying a banquet. The
DOCTOR enters. MACBETH goes to him.*

MACBETH
[whisper] Is Banquo dispatched?

DOCTOR

[gruff Glaswegian] My Lord, his throat is cut, that I did for
him. *[mimes, making 'quack' sound]*

MACBETH

Thou art the best o'th'cut-throats; And tell me
Thou didst the like for Fleance?

DOCTOR

Aye, I did.
Safe with his father in a ditch he bides.

MACBETH

Now get thee gone.

LENOX

[calls to MACBETH] May it please your Highness sit?

MACBETH

Here had we now our country's honour roof'd,
Were the grace'd person of our Banquo present.

DOCTOR

[whisper] Now, Banquo!

*BANQUO appears, his face white with chalk, wearing
the perception filter necklace. He sits in MACBETH's
seat. MACBETH reacts with horror at the sight.*

MACBETH

The table's full!

LENOX

*[indicates where BANQUO is sitting, but not seeing
BANQUO]* Here is a place reserv'd, Sir.

MACBETH

Where?

LENOX

Here, my good Lord. What is't that moves your Highness?
And makest thou tremble and wide of eye?

MACBETH

[points to BANQUO] Prithee, see there!
Behold! Look! Lo! How say you?
If charnel-houses and our graves must send
Those that we bury back to haunt us?

DOCTOR

[whisper] Banquo, now!

> *BANQUO adjusts the necklace and vanishes.*
> *No one has seen him apart from MACBETH.*

LADY MACBETH

What is't that hast unmann'd thee?

MACBETH

If I stand here, I saw him. Banquo!

LADY MACBETH

Fie! For shame!
My worthy Lord, your noble friends do lack you.

MACBETH

[regaining his composure] I do forget, –
Do not muse at me, my most worthy friends,
I have a… strange infirmity, which is nothing
To those that know me. Come, love and health to all;
Then I'll sit down. *[he gingerly sits in the empty seat]* Give me
some wine: fill full;

Drink to our dear friend Banquo, whom we miss;
Would he were here!

DOCTOR
[whisper] Now!

BANQUO reappears, standing on the table in front of MACBETH.
MACBETH leaps out of his seat in terror, spilling his wine.

MACBETH
[to BANQUO] Avaunt! And quit my sight! Let the earth hide
thee!
Thy bones are marrowless, thy blood is cold;
Thou hast no speculation in those eyes,
Which thou dost glare with. Hence, horrible shadow!
Unreal mock'ry, hence!

BANQUO adjusts the necklace and vanishes.
MACBETH shakes in fright. LADY MACBETH
approaches him.

LADY MACBETH
You have displac'd the mirth, broke the good meeting,
With most direful disorder.

MACBETH
Can such things be?
My skin is blanch'd with fear.

ROSSE
What things, my Lord?

LADY MACBETH
I pray you, speak not; he grows worse and worse;
But go at once.

LENOX
>Good night, and better health
>Attend his Majesty!

LADY MACBETH
>A kind good night to all!

>*ROSSE, LENOX, the Lords and Attendants leave.*

DOCTOR
>*[whisper]* All right, Banquo. I think we have convinc'd
>Him of your death. You must to Wales depart
>And join thy son.

BANQUO
>*[invisible, whisper]* Yes, Doctor. It's been fun.

>*MACBETH stands aside, deep in thought.*

MACBETH
>I will go to consult the Weird Sisters:
>More shall they speak; for now I am bent to know
>By the worst means, the worst. I am in blood
>Stepp'd in so far, that, should I wade no more,
>Returning were as tedious as go o'er.
>Strange things I have in head, that will to hand,
>Which must be acted, ere they may be scann'd!

>*Exit MACBETH and LADY MACBETH.*

DOCTOR
>'Weird Sisters'? Oh my word, he means Jamie, Zoe and me!

>*In a sudden panic, DOCTOR scurries out.*

ACT IV, SCENE I – A CAVE ON THE HEATH

The DOCTOR, JAMIE and ZOE stand around a bubbling cauldron. They are all wearing rags and apply mud to their faces to disguise themselves as witches.

JAMIE

I still don't know why we have to do this.

ZOE

Because it's in the play.

DOCTOR

And not just that
We have to lull Macbeth into a false
Sense of security so when Macduff's
Army arrives it catches him off-guard.

JAMIE

What's yon gadget you have there, Doctor?

He indicates a large, metal box set to one side of the cave.

DOCTOR

An instrument of mine that should convince
Macbeth of evr'y word that we shall speak.
Are we all set?

ZOE

Of course. I've memorised
The entire play by heart.

DOCTOR

Then you can do
The talking. Jamie and I will join in.

JAMIE

Join in with what? *[Sound of someone approaching through undergrowth]*

ZOE

[clears throat, then clearly and precisely] By the pricking of my thumbs,
Something wicked this way comes.

Enter MACBETH.

MACBETH

How now, you secret, black, and midnight hags!
What is't you do?

ZOE

A deed without a name. *[DOCTOR and JAMIE join in, doing women's voices]*

MACBETH

I conjure you, by that which you profess,
Howe'er you come to know it, answer me:
Though you untie the winds and let them fight
Against the churches; though the yesty waves –

DOCTOR

    [high-pitched] Yes, yes, yes, you can skip all that.

ZOE

    We'll answer. Pour in sow's blood, that hath eaten
    Her nine farrow; grease that's sweaten
    From the murderer's gibbet throw
    Into the flame. *[Adds ingredients to cauldron, causing a burst
    of flame]*

ZOE, DOCTOR, JAMIE

    Come, high or low;
    Thyself and office deftly show!

      *The Doctor clicks a device. The metal box doth project an
         image of Macduff on the cave wall.*

ZOE

    [speaking into a wand, her voice is amplified and echoes terribly]
    Macbeth! Macbeth! Macbeth! Beware Macduff!
    Beware the Thane of Fife. Dismiss me. Enough!

MACBETH

    Whate'er thou art, for thy good caution, thanks;
    Thou hast harp'd my fear aright: but one
    word more –

ZOE

    He will not be commanded: here's another,
    More potent than the first.

      An image of a newborn baby appears on the cave wall.

ZOE

    Be bloody, bold, and resolute; laugh to scorn
    The power of man, for none of woman born
    Shall harm Macbeth.

MACBETH

> Then live, Macduff: what need I fear of thee?
> But yet I'll make assurance double sure,
> And take a bond of fate: thou shalt not live.

Now the image of a boy wearing a crown appears.

MACBETH

> What is this
> That rises like the issue of a King,
> And wears upon his baby-brow the round
> And top of sovereignty?

ZOE

> Be lion-mettled, proud; and take no care
> Who chafes, who frets, or where conspirers are:
> Macbeth shall never vanquish'd be until
> Great Birnam wood to high Dunsinane hill
> Shall come against him.

The Doctor waves his hand and the image vanishes.

MACBETH

> That will never be!
> Who can impress the forest, bid the tree
> Unfix his earth-bound root? And yet my heart
> Throbs to know one thing: tell me, if your art
> Can tell so much: shall Banquo's issue ever
> Reign in this kingdom?

ZOE

> Seek to know no more!

MACBETH

> I will be satisfied: deny me this,
> And an eternal curse fall on you! Let me know!

DOCTOR
[high-pitched] Show!

JAMIE
[high-pitched] Show!

ZOE
Show his eyes, and grieve his heart;
Come like shadows, so depart! *[she throws another ingredient into the cauldron. There is a burst of flame and it is extinguished]*

> *The DOCTOR works his magick and images of future Scottish Kings appear on the cave wall. MACBETH is transfixed.*

MACBETH
Thou art too like the spirit of Banquo: down!
Thy crown does sear mine eye-balls. And thy hair,
Thou other gold-bound brow, is like the first.
A third is like the former. Filthy hags!
Why do you show me this? A fourth! Start, eyes!
What, will the line stretch out to the crack of doom?

> *The magick light show ends, plunging the cave into darkness.*

MACBETH
Where are they? Gone? Let this pernicious hour
Stand aye accursed in the calendar!
Come in, without there!

> *Enter LENOX, carrying a torch.*

LENOX
What's your grace's will?

MACBETH

Saw you the Weird Sisters?

LENOX

No, my lord.

MACBETH

Infected be the air whereon they ride;
And damn'd all those that trust them! I did hear
The galloping of horse: who was't came by?

LENOX

'Tis two or three, my lord, that bring you word
Macduff is fled to England.

MACBETH

Fled to England?

LENOX

Ay, my good lord.

MACBETH

Time, thou anticipatest my dread exploits:
The castle of Macduff I will surprise;
Seize upon Fife; give to the edge o' the sword
His wife, his babes, and all unfortunate souls
That trace him in a line.

Exeunt MACBETH and LENOX

*The DOCTOR, JAMIE and ZOE emerge from their
hiding place.*

JAMIE

He's going to have Macduff's wife killed? And his
children too?

DOCTOR

> Not if we can help it, Jamie. We shall
> Change from these rags to villains' semblances.
> And once more offer our assistance to the King…

ACT IV, SCENE II – A ROOM IN MACDUFF'S CASTLE

LADY MACDUFF and her children. The DOCTOR and JAMIE enter, dressed in their hooded coats.

LADY MACDUFF

> Who art thou? To me thou art not known.

DOCTOR

> We've come to warn thou art in danger grave,
> If you will take a homely man's advice
> Be not found here; hence, with thy little ones.

LADY MACDUFF

> Whither should I fly? I have done no harm.

DOCTOR

> The King's accus'd Macduff of treason high,
> And ordered us to kill you here to-night.

LADY MACDUFF

> Thou liest, thou… shag-hair'd villain!

JAMIE

> But if you leave now, we'll go back to the King and *tell* him we killed you,
> And he won't send anyone else after you.

DOCTOR
Alas, I fear you have no choice, my dear.
When I the word 'run' sayeth, thou must run!
Run!

Exit LADY MACDUFF with her children.

JAMIE
Do you think they'll be all right, Doctor?

DOCTOR
Oh, I expect so Jamie. They just need
To lie low for a while. You see, Macbeth's
Black days are number'd. Come, let's Zoe find.

ACT V, SCENE I – A PASSAGE IN MACBETH'S CASTLE

The DOCTOR and JAMIE enter, still in their coats.
A WAITING GENTLEWOMAN enters.

GENTLEWOMAN
Pray, is one of you the Doctor?

JAMIE
The Doctor?

GENTLEWOMAN
The Doctor call'd to discover the cure
To Lady Macbeth's new-found malady.

The Doctor removes his coat.

DOCTOR
[adopts educated Edinburgh accent] Yes, I'm the Doctor.
[rubs hands] Tell me, what is wrong?

GENTLEWOMAN

Since his majesty went into the field, I have seen
her rise from her bed, throw her night-gown upon
her, unlock her closet, take forth paper, fold it,
write upon't, read it, afterwards seal it, and again
return to bed; yet all this while in a most fast sleep.

DOCTOR

I see… In this slumbery agitation, besides her
Walking and other actual performances, what, at any
Time, have you heard her say?

GENTLEWOMAN

That, sir, which I will not report after her.

DOCTOR

You may to me: and 'tis most meet you should.

Enter LADY MACBETH holding a lit candle.

GENTLEWOMAN

Lo you, here she comes! This is her very guise;
and, upon my life, fast asleep. Observe her; stand close.

JAMIE

Doctor, her eyes are open!

GENTLEWOMAN

Aye, but their sense is shut.

*LADY MACBETH puts down the candle
and wrings her hands.*

DOCTOR

What is it she does now? Look, how she rubs her hands.

GENTLEWOMAN

It is an accustomed action with her, to seem thus
washing her hands: I have known her continue in
this a quarter of an hour.

DOCTOR

She would appear to be beset with some
Form of obsessive compulsive sickness…

LADY MACBETH

Out, damned spot! out, I say!   One: two; why,
then, 'tis time to do't. – Hell is murky! – Fie, my
lord, fie! Yet who would have thought the old man
to have had so much blood in him?

DOCTOR

Did you hear that?

GENTLEWOMAN

She has spoke what she should not, I am sure of
that: Heaven knows what she has known.

Exit LADY MACBETH.

DOCTOR

Will she go now to bed?

GENTLEWOMAN

Directly.

DOCTOR

Foul whisperings are abroad: unnatural deeds
Do breed unnatural troubles. Look after her;
Remove from her the means of all annoyance,
And still keep eyes upon her. So, good night.

GENTLEWOMAN

> Good night, good doctor.

> *Exit GENTLEWOMAN.*

JAMIE

> What d'you think's wrong with her, Doctor? Sleepwalking?

DOCTOR

> I'm not sure, Jamie. Parasomnia?
> Or post-traumatic stress disorder? Poor woman.

JAMIE

> Poor woman? She goaded her husband into killing the King!

DOCTOR

> Even so. But I'm afraid she is past curing now...

ACT V, SCENE III – THE CASTLE THRONE ROOM

Enter MACBETH with the DOCTOR and JAMIE,
no longer wearing their coats. The DOCTOR is wearing
spectacles. A SERVANT enters.

MACBETH

> The devil damn thee black, thou cream-faced loon!
> Where got'st thou that goose look?

SERVANT

> There is ten thousand –

MACBETH

> Geese, villain?

SERVANT

> Soldiers, sir. Outside the castle!

MACBETH

What soldiers, whey-face?

SERVANT

The English force, so please you.

MACBETH

Take thy face hence.

Exit SERVANT

How does your patient, Doctor?

DOCTOR

[educated Edinburgh accent] Not so sick, my lord,
As she is troubled with thick coming fancies,
That keep her from her rest.

MACBETH

Cure her of that.
Canst thou not minister to a mind diseased,
Pluck from the memory a rooted sorrow,
Raze out the written troubles of the brain
And with some sweet oblivious antidote
Cleanse the stuff'd bosom of that perilous stuff
Which weighs upon the heart?

DOCTOR

Therein the patient
Must minister to himself.

MACBETH

Throw physic to the dogs; I'll none of it. –
I'll put mine armour on and fetch my staff.
I will not be afraid of death and bane,
Till Birnam forest come to Dunsinane.

Exit MACBETH.

JAMIE

He's going to fight the English army?

DOCTOR

Yes, fight and lose. He won't realise his men
Are massively outnumber'd 'til it's far
Too late because the English army have
Their awful might conceal'd with camouflage.

JAMIE

Camouflage?

DOCTOR

With leaves and twigs disguis'd. So as far as
Macbeth's concern'd, it will the prophecy fulfil
As Birnhamwood has come to Dunsinane.

JAMIE

Well, not really.

DOCTOR

It's close enough. Let's go and Zoe meet.
Our time in Scotland's discord is complete.

ACT V, SCENE IV – BIRNAM WOOD

*Drum and colours. Enter MALCOLM, SIWARD and
BANQUO, MACDUFF, MENTEITH, LENOX,
ROSSE, and the Soldiers of the English army.*

SIWARD

What wood is this before us?

MENTEITH

The wood of Birnam.

Enter ZOE.

MACDUFF
> What art thou doing here? This is no place
> For thou, wee slip o'girl. Get thee off home!

ZOE
> I think you'll find I can for myself fend.

BANQUO
> I recognise this girl! She is a friend,
> Who aided my escape and of my son.

ZOE
> How is he?

BANQUO
> In Wales he's safe bestow'd. I left to join
> This army rais'd by Malcolm to remove
> Macbeth the cruel usurper from his throne.

ZOE
> That's why I'm here. I've come to help you win.

MALCOLM
> Help us? How?

ZOE
> Macbeth expects your assault. So you must
> Use camouflage.

MACDUFF
> What is this 'camouflage'?

ZOE
> The wood. Let every soldier hew down a bough,
> And bear't before him: thereby shall you shadow
> The numbers of our host, and make discovery
> Err in report of you.

MALCOLM

It shall be done.

The soldiers disguise themselves with branches and leaves
and exeunt, marching.

ACT V, SCENE VIII – OUTSIDE THE CASTLE

The DOCTOR, JAMIE and ZOE slip outside through
a side-door as the English army lays siege to main gates.

JAMIE

Those gates won't keep them out for long.

DOCTOR

And that's it for Macbeth. Out, out, brief candle.
Life's but a walking shadow, a poor player
That struts and frets his hour upon a stage
And then is heard no more.

The English army smash through the gates and swarm
into the castle.

ZOE

So Shakespeare had it right, then, after all?

DOCTOR

Not quite.
Imagine if we had not interven'd
How many more foul murders there'd have been.

They approach a blue dwelling.

JAMIE

Aye, but apart from that, it's all like in yon book.

DOCTOR

Yes. History is on its proper course
Once more. I must remember to make sure
That Shakespeare writes it straight, next time we meet.

ZOE

Next time? Next time? You've met Shakespeare before?

The DOCTOR opens the door to the dwelling and enters.

JAMIE

Aye. Who is this Shakespeare fella, anyway? I've never heard of him.

ZOE

[entering] Oh, Jamie, you have so much left to learn.

JAMIE

[entering] Hey, at least I know my own history.

DOCTOR

[entering] That's what began our troubles, don't forget…

*Trumpet. Wheezing, groaning. The DWELLING vanishes
– leaving the DOCTOR, JAMIE and ZOE standing
where it once stood.*

JAMIE & ZOE

Oh, *Doctor!*

*Wheezing, groaning. And then the DOCTOR, JAMIE
and ZOE vanish, bound for another adventure.*

Friends, Daleks, Cybermen…

CYMBELINE

Another early draft of a scene, this time from the play Cymbeline.

While in Rome, Posthumus, son-in-law to Cymbeline, King of the Britons, accepts a drunken wager from an Italian nobleman, Iachimo. Iachimo proposes to test the loyalty of Imogen, daughter of Cymbeline and wife to Posthumus. He travels to the British city of Lud's Town, where he asks Imogen to guard a box of valuables. Imogen takes the box into her chamber. Iachimo has concealed himself inside it, and plans to emerge at midnight to collect evidence that he has spent time in her room: notes on the furnishings, and the bracelet that Imogen always wears – a love token from Posthumus.

<u>SCENE II</u>

Imogen's bedchamber in Cymbeline's palace: a trunk in one corner of it. IMOGEN in bed, reading; a Lady attending

IMOGEN
    Who's there? My woman Helen?

LADY
    Please you, madam.

IMOGEN
    What hour is it?

LADY
    Almost midnight, madam.

IMOGEN
    I have read three hours, then: mine eyes are weak:
    Fold down the leaf where I have left: to bed.
    What is't, Helen? Fold down the page, I say.

LADY

Excuse the fancy, madam.

IMOGEN

Speak, Helen.

LADY

'Twas but a shadow dancing on the wall. The light, madam.
Fitfully it burns.

IMOGEN

Take not away the taper, leave it burning;
And if thou canst awake by four o' the clock,
I prithee, call me. Sleep hath seized me wholly.

Exit Lady

To your protection I commend me, gods.
From fairies and the tempters of the night
Guard me, beseech ye.

Sleeps. IACHIMO comes from the trunk

IACHIMO

The crickets sing, and man's o'er-labour'd sense
Repairs itself by rest. Our Tarquin thus
Did softly press the rushes, ere he waken'd
The chastity he wounded. Cytherea,
How bravely thou becomest thy bed, fresh lily,
And whiter than the sheets! That I might touch!
But kiss; one kiss!

DOCTOR comes from the trunk.

IACHIMO

But soft! Who's this?

DOCTOR

> I would not do that deed if I were you.
> There is a word for chaps who play such tricks
> And judges say it in a court of law.

IACHIMO

> The Doctor who attends the Britons' Queen!
> Concealed within the trunk! How is this so?
> I saw you not when I was 'prisoned there.

DOCTOR

> I know about your filthy little game;
> Your plot to thieve the bracelet from her wrist.
> This wager that you made with Posthumus
> To test the faith of Lady Imogen
> Shames you and he, not her. I don't suppose
> You've antiseptic hand-rub in that purse?

IACHIMO

> I prithee, gentle Doctor, hold thy peace.
> She will awake and then we both are lost.

DOCTOR

> Oh are we, now? Tell me, Iachimo –
> Iachimo – I like the ring of that –
> 'Tis very like a word I like to cry
> When jumping into dangers unexplored –
> Tell me who sleeps before us in this room.
> What do you see?

IACHIMO

> Rubies unparagon'd,
> How dearly they do't! 'Tis her breathing that
> Perfumes the chamber thus: the flame o' the taper
> Bows toward her, and would under-peep her lids,
> To see the enclosed lights, now canopied

Under these windows, white and azure laced
With blue of heaven's own tinct.

DOCTOR

Blimey, Iachimo, you do go on.
And now you're taking notes. You're quite
The thorough villain.

IACHIMO

I will write all down:
Such and such pictures; there the window; such
The adornment of her bed; the arras; figures,
Why, such and such; and the contents o' the story.

DOCTOR

What story do you mean, Iachimo?

IACHIMO

There o'the wall, Doctor. It is the tale
Of Tereus the tyrant King of Thrace
And the unhappy sister to his wife
Told in a tapestry. See, here she is,
Fair Philomel, before the gods assuaged
Her mortal pain by strange transformation
Into the bird that makes its song by night
I'faith! What's here? What mockery is this?
There is but woven air where she once stood.

DOCTOR

Ah, that's not good. In fact, it's very bad.
I saw this happen on Trafalgar Square –
Or under it, at least – and then the world
Was nearly blown to pieces that same day.
Perhaps you ought to climb back in the box
Close up the lid; pray for deliverance.

IACHIMO

> Hide in a cask, affrighted by a void
> On an arras? My role is not yet played.
> My wager lies unwon. Here is the prize:
> The band of treasure wrapp'd about her wrist.

DOCTOR

> I really would stand back, Iachimo!

IACHIMO

> O sleep, thou ape of death, lie dull upon her!
> And be her sense but as a monument,
> Thus in a chapel lying! Come off, come off:
> (Taking off her bracelet)
> As slippery as the Gordian knot was hard!
> 'Tis mine; and this will witness outwardly,
> As strongly as the conscience does within,
> To the madding of her lord.

IMOGEN

> Who is't? What ho?

DOCTOR

> His name's Iachimo. He's come from Rome.
> In Italy. The country like a boot.

IACHIMO

> My lady, I profess with all my heart –

IMOGEN

> Your heart, methinks is not so bounteous.
> Why stand you here within my private rooms?

IACHIMO

> Into this chamber, madam, we intrude.
> Though in the royal court of Cymbeline
> It seems we left our manners at the door.

DOCTOR

But so did you, I think, fair Imogen.

IACHIMO

Sweet doctor hold thy peace. Keep thou this thought:
This lady holds us both within her power.

DOCTOR

She's no lady. She's not even from earth.
Her habitation was a torrid world
Of seas and lakes and swamps wherein which dwelt
Great milky worms they called the Skaruwen.
Those roaring beasts that were all boiled like eels
When stellar blasts screamed through her tract of space.
Is that not right?

IMOGEN

Thou speak'st a bitter truth.
And now this jay of Italy has robbed
The bracelet from my arm, I am
Obliged to cast this form into the air
And bid adieu to honest Imogen.

IACHIMO

> God's wounds! Mine eyes! She melts into the air.
> Where is the daughter of the Britons' king?
> The lady Imogen hath not the mien
> Of some abomination dredged up from the deep.
> What is this creature stands before us now?
> Some great soused gurnet upright o'the land,
> Skin like calvered salmon, all o'erwrought
> With crimson veins and stinking barnacles.
> Two eyes like coals new fallen from the fire,
> Which even now are burning 'pon my skin.
> Doctor, prepare some draught that might relieve
> The sickness that has overpowered my sight.

DOCTOR

> I have no herb for this, Iachimo.

IACHIMO

> Release me, sir, from my delusions!

DOCTOR

> You do not dream. There is no spell to break.

IACHIMO

> O lamprey, spare me. Hath I offended thee?

IMOGEN

> The paltry human thing begs for his life.
> He would not beg if he could see my mind;
> The fortune that I conjure for his world:
> Italy drowned, the world a brackish lake,
> Rank weeds splitting the stones of old Lud's town;
> Hot steam above the Thames; and there,
> Pushing its way through river's warm expanse,
> The fatal body of the Skarasen.
> Answering its mistress' royal call.

DOCTOR
You lot are fixed on feudal circumstance.
Dukedoms, lairdships, empires and the great throne
Of Tulloch's Golden Haggis Lucky Dip.
Iachimo, may I toss this your way?
Within this little chamber, as you see
A greater war is just about to bloom.
Remember I said get back in that box?
It might be time to think on this again.

IACHIMO
To the trunk again, and shut the spring of it.
Swift, swift, you dragons of the night, that dawning
May bare the raven's eye! I lodge in fear;
Though this was heavenly angel, hell is here.

Clock strikes

One, two, three: time, time!

Goes into the trunk.

DOCTOR
Adieu, Iachimo. Remain inside,
Crouched on your frowsty bed of Roman coin.

IMOGEN
'Tis best he does not see the final act.
The next scene, gentle Doctor, is our own.
When it concludes, Britain shall be my realm.

The scene closes

Blow Time Winds and crack reality.

DIARY EXTRACT

The following extract, apparently transcribed into the Notebooks from William Shakespeare's personal diary for late 1601 (exact date not recorded) describes a meeting to discuss plans for the play that became Twelfth Night. *Richard Burbage was the leading player of the Lord Chamberlain's Men, and (together with his older non-acting brother Cuthbert) owned the Globe theatre where they performed.*

The identity of the third member of the group – referred to in Shakespeare's notes only as 'He' or 'that man' – has never been established. The implication is of course obvious from the fact that Shakespeare included it in the Notebooks.

Finally, Burbage did agree to discuss my outline for the play which we hope to perform come Candlemas, and so we retired to a nearby hostelry thinking ourselves to be well received and accorded some privacy.

But He was there again. Methinks that man doth follow me like a shadow some times. He is everywhere, and always where least expected. Though I must confess His company is agreeable, His conversation witty, and His insights oft-times display a depth of perspicacity that belies His outward demeanour. Burbage seemed content to allow Him to join our table, and who was I to offer objection.

'Why wear you celery on your coat?' Burbage asked as we awaited our ale. He is never one to hold back when he could speak his mind.

'I think it's rather fashionable, don't you?' He replied. Seeing Burbage's perplexity, His expression changed. 'Oh, you don't. Well, never mind.' And so He turned to me. 'So what's this next play all about?'

I made a study of contemplation, as if gathering my thoughts, and then explained, 'I have a mind to write of a woman who is shipwrecked. Fearing her brother is lost at sea and anxious for her own safety, she doth counterfeit her own brother in character and apparel. I draw inspiration,' I confided, 'from the story "Of Apollonius and Silla" by one Barnabe Rich.'

He made a face at this, and not by cause of sour ale. 'I prefer the original. By Matteo Bandello. I met him once, you know. Dreadful table manners.' He glanced at Burbage who was quaffing his ale with gusto. 'Not a problem here, I see.'

'So, what is the plot?' Burbage asked, belching vapours and wiping his mouth with his hand.

'The woman, who I shall name Viola, is therefore dressed as a man…'

'Always gets laughs, that,' Burbage said.

I pressed on. 'She is employed to carry favours of love between two estranged lovers. And doing so, she her self doth fall in love with the man. And the woman doth fall in love with Viola, thinking her to be a man also.' I let them consider this, and continued. 'I have in mind also a grotesque. A great fat man fond of ale, who is in conflict with the upright servant of the woman in love. His name shall be –' I paused to think of a suitable nomenclature. In the silence, Burbage again gave vent to the wind, and bestowed upon me the very name I sought. 'Belch. Sir Toby Belch. There will be great bating and amusement. The fat man, Belch, will make it seem that the officious servant is mad for love of his mistress. And there is another too, a coward who is coerced into duelling.'

Burbage clapped his hands at this. 'And a fool. There must be a fool.'

'Assuredly,' I told him.

'Um, sorry,' He said. 'But, well, that all sounds rather complicated to me. Don't you think?'

'Complicated?'

'Yes. And it doesn't make much sense either, does it? I mean who'd mistake a woman dressed as a man for the real thing?'

'She will be played,' Burbage pointed out, 'by a boy.'

He frowned at this, lifted his ale, then replaced it without drinking. 'Sorry, it's probably me, but you will present a boy playing a woman who dresses as a man who then falls in love with a man who I assume is really a man, but while still pretending to be a man. And then the woman the man loves, who is also I assume in reality a boy, falls in love with the man who is actually a woman played by a boy pretending to be a man.'

'Pretending to be her own brother,' I added, by way of clarification. 'You have it exactly.'

'And who is the brother pretending to be?' He asked.

'Why no one. He is himself.'

He shook his head. 'Still not sure,' He admitted. 'So what happens when the real brother turns up? If he does turn up.'

'He must,' Burbage agreed.

'Then won't that be confusing? For the audience but also for all the characters. What if someone meets the brother and thinks it's the boy being a girl who's pretending to be the brother but is actually the sister?'

'That must happen too!' Burbage exclaimed, slamming down his empty tankard. 'Confusion, thy name is...' He broke off. 'Confused,' he decided, signalling for another draught.

'I think you've had quite enough for today,' He said, removing Burbage's tankard from out of reach.

As could be predicted, Burbage was all of a rage at this. 'Dost thou think because thou art virtuous there shall be no more cakes and ale?'

I made a small note of this on my papers. And also of His response – 'Oh the whirligig of time,' He murmured.

'What's it called, this comedy of yours?' Burbage demanded. 'It is a comedy?'

'It is a comedy,' I assured him. 'And I have decided to name it "Twelfth Night".'

There was pause at this as both the company made contemplation of the title.

'"Twelfth Night"?' Burbage said at last. '"Twelfth Night".'

'Why?' He asked – ever my buggbear.

'Forsooth, because the play doth explore the misrule of the fat man, Belch.'

'Fair enough, But why "Twelfth Night"?'

'There shall be songs and mummery and riotous disorder,' I told Him, my exasperation growing by the moment.

'No, I still don't see what "Twelfth Night" has to do with a love triangle between a man, a woman, and a cross-dressing woman played by a boy.'

Feeling my blood begin to boil at this, I left Him and Burbage to debate the matter. '"Twelfth Night" is the name I have given my play,' I told them with authoritie as I took my leave. 'But I am merely the playwright, and you may call it what you will.'

THE DREAM

One of the strangest sections of the Shakespeare Notebooks comprises a form of 'dream diary', in which he apparently recorded characters, situations and images that had come to him in his sleep.

Of particular interest is the entry for 24 June 1594 – the midsummer night's dream that inspired one of his well-loved plays, which appears to have been born entirely from his imagination…

Last night, I had the most wonderful and strange vision, of foreign worlds and marvellous beasts, in the most fantastical comedy. I must set it down before the memory fades. It may make, with some changes, a most excellent play. I must also remember not to drink the ale at the Elephant again as my head feels like it's had an elephant sitting on it.

A STRANGE WOODLAND

Sandy ground. Bubbling streams. Low mist.
The trunks of immense bamboo trees stretch up
to a night sky garlanded with stars.

Enter, from one side, the BUTTERFLY KING, riding
upon a GIANT ANT accompanied by GIANT GRUBS
and WOODLICE-MEN. And on the other side, the
BUTTERFLY QUEEN on another GIANT ANT
accompanied by BUTTERFLY CHILDREN.

BUTTERFLY KING
  Ill met by moons' light, proud Tetynia.
  Wherefore comest thou to the flower forest?

BUTTERFLY QUEEN
>How now, Hrobron! I come from my temple
>Of light upon the moon Taron to warn
>Thou of a fresh contagion on our world.
>A ship has fallen from the stars and is
>Not far remov'd.

BUTTERFLY KING
>Within this wood, you say?

BUTTERFLY QUEEN
>If thou wouldst please accompany my flight
>I'll mark you where they lie. They are most strange;
>Nasty, brutish and short.

BUTTERFLY KING
>I'll go with thee.

BUTTERFLY QUEEN
>Then follow close behind and thou wilt see
>Their form and overhear their conference.

The BUTTERFLY KING and BUTTERFLY QUEEN
join hands and fly off together.

ANOTHER PART OF THE WOOD

A group of WAR-GOBLINS enter,
led by FIELD MAJOR KRYNTZ.

KRYNTZ
>This will make a marvellous convenient location for our
>preparations. Is all our division met?

STOMBOT
> I recommend you call them gen'rally according to the
> roll-call.

KRYNTZ
> Here is the list of every trooper that's been deemed fit to
> participate in our mission; a dramatic reconstruction to be
> performed before the Group Marshal and the latest batch
> of clone recruits at the victory ceremony in celebration of
> our glorious conquest of the Isop galaxy and the merciless
> subjugation of its many puny species.

STOMBOT
> Field Major Kryntz –

KRYNTZ
> What sayest thou, Fabricator Stombot?

STOMBOT
> Why have we been charged to enact a reconstruction?
> Why not just show the high-definition holovisual
> recording?

KRYNTZ
> A recording does not possess the same dramatic impact
> as a live reconstruction. We are to interpret the incidents
> according to the traditions of our glorious Sontaran culture.

STOMBOT
> What incidents are we to reconstruct for the imperial
> leader?

KRYNTZ
> Our reconstruction is, "The Most Glorious Defeat and Most
> Deserv'd Death of the Trifling Rutan Foe at Fang Rock.".

STOMBOT

Was that victory not achieved by an inferior race without our involvement?

KRYNTZ

Exactly, Stombot. The reconstruction is to demonstrate that the Rutan foe are so feeble they can be defeated by a puny subspecies without our active participation.

STOMBOT

But if that is the case, why have we not defeated them?

KRYNTZ

That is seditious talk, Stombot! As you know our final victory over the Rutan is imminent and has been imminent for the last one hundred thousand years. Now, answer as you are called. Fabricator Stombot?

STOMBOT

Sontar-ha! Name what part I am for, and proceed.

KRYNTZ

You are set down for the Doctor.

STOMBOT

What is 'the Doctor'? A heroic warrior or a cruel tyrant?

KRYNTZ

A most heroic warrior that shows no mercy to the Rutan!

STOMBOT

A most fitting role. I assure you I will be absolutely pitiful.

KRYNTZ

You must not be pitiful, you must be pitiless, or we will all be reassigned to latrine maintenance duties. Ventilation Engineer Flaxis?

FLAXIS
Sontar-ha!

KRYNTZ
Flaxis, you must take the part of Leela.

FLAXIS
What is Leela? Another heroic warrior?

KRYNTZ
He is the Doctor's boy assistant who taunts the Rutan during its lingering death throes. Uniform Fitter Starvel?

STARVEL
Sontar-ha!

KRYNTZ
You shall play Vince 'the light-house keeper'. Utensil Technician Stoun?

STOUN
Sontar-ha!

KRYNTZ
You, 'Lord Palmerdale'. Myself, 'Colonel Skinsale'. Snarg the Construction Operative, you the Rutan's part.

SNARG
Have you the Rutan's part written? Pray you, if it be, give it me, for I am slow of study.

KRYNTZ
You may do it extempore, for it is nothing but dying.

STOMBOT
Let me play the Rutan too: I will play it so well, I will convince everyone watching I am the genuine article!

KRYNTZ

And set off an intruder alert and have us all immediately vapourised! No. You must play no part but the Doctor.

STOMBOT

Then you must write a prologue; let the prologue say, for the more better assurance, that Snarg is not a Rutan, but Snarg the Construction Operative; this will avoid any confusion.

KRYNTZ

Well it shall be so. But there is two hard things; that is, to bring the lighthouse into the imperial chamber; for, you know, the Rutan scout was destroyed by the explosion of a lighthouse. So one of us must come in with a mound of rocks and a lantern, and say he comes to represent the person of 'Lighthouse'. Then there is another thing: we must have a Rutan scoutship in the imperial chamber; for the story includes a Rutan scoutship crashing.

STOUN

You can never have a Rutan scoutship crashing in the imperial chamber! What say you, Stombot?

STOMBOT

Some man or other must present 'Scoutship': and let him have some glass about him to signify crystalline infrastructure!

KRYNTZ hands out scripts.

KRYNTZ

If that may be, then all is well. Let us begin: Here are your scripts. When you have spoken your speech, Stombot, enter that cave: and so every Sontaran according to his cue.

Enter the WHITE-HAIRED OLD DOCTOR,
crouched behind a rock.

OLD DOCTOR
    O my goodness!
    What thick-set ruffians have we swagg'ring here,
    And disrupting the harmony o'Vortis?
    It will not do, it will not do at all! *[chuckles]*

He scurries into a nearby cave.

KRYNTZ
    Speak, Stombot. Flaxis, stand forth.

STOMBOT
    [reads] "Where is this place we have arriv'd, it is not Britain."

KRYNTZ
    Say "Bright-ton", not "Brit-ain"!

FLAXIS
    [reads] "Doctor, has your inferior travel capsule failed
    again?"

STOMBOT

"Yes, the localised atmospheric conditions caused a
navigation error in my inferior technology."

FLAXIS

"What is that over yonder?" *[points]*

STOMBOT

"'Tis a light-house. Come, young boy. We must investigate
and destroy any Rutans that have gained access."

KRYNTZ

Now you must enter into the cave with Fraxis as we cut to
inside the lighthouse.

STOMBOT and FLAXIS enter the cave.

KRYNTZ

Uniform Fitter Starvel, that is your cue. Stand forth!

STARVEL

[reads] "I am Vince, commander of the light-house. What is
that I see – a falling star. It goes under the sea and doth glow
most mysteriously. I shall disregard it as a meteor as I am a
foolish primitive incapable of recognising a Rutan scoutship.
And what is this? A mist –"

FLAXIS runs from the cave.

FLAXIS

[breathless] Field Major Kryntz, we must evacuate!
At once!

KRYNTZ

What is it?

FLAXIS

Fabricator Stombot isn't Fabricator Stombot!

*STOMBOT emerges from the cave. But transformed into
a GREEN-EYED MONSTER with numerous tentacles!*

KRYNTZ

O monstrous! O strange!

GREEN-EYED MONSTER

What's the matter? I am Fabricator Stombot of the fourth
Sontaran Army Space Corps. Why do you stare at me thus?

STOUN

Thou art transform'd! Thou art a *Rutan*!

GREEN-EYED MONSTER

What? *[sighs]* Oh no. Our metamorphic field generator must
have failed.

KRYNTZ

It must have eliminated Fabricator Stombot and taken his
place. But why? To gain access to the victory ceremony and
assassinate the Group Marshall!

GREEN-EYED MONSTER

You have reason'd correctly. And we would have succeeded
had our disguise not destabilised. Now you must all be
destroy'd! Miserable Sontaran rabble!

*It starts crackling with lightning.
The WAR-GOBLINS back away.*

KRYNTZ

All troopers! We must leave this planet at once! Our mission
has been compromised! Abandon Vortis! Repeat! Abandon
Vortis!

Exeunt WAR-GOBLINS pursued by
GREEN-EYED MONSTER.

The WHITE-HAIRED OLD DOCTOR
emerges from the cave.

OLD DOCTOR
    [giggling] A simple matter to deactivate
    The Rutan's metamorphic field with my
    Reacting Collator! I doubt that we
    Will see them hence, not for awhile at least!
    Good riddance to bad rubbish, I should say!
    Now where did I leave Steven and Dodo?

Exit the WHITE-HAIRED OLD DOCTOR.

> *The time is out of joint, oh cursed spite,*
> *why can't the TARDIS ever get it right?*

A PROLOGUE

This fragment appears to be a variant version of the opening of Henry
V, *Shakespeare's play about the events before and after the battle
of Agincourt in 1415. The Chorus is the first character to appear,
and explains to the audience that the lifelike depiction of such a
conflict is beyond the means of a little group of actors in a wooden
theatre. However, the battle described here is fought by agents other
than the French and the English – combatants such as the Daleks,
the Nightmare Child and the Time Lords are mentioned. This war
would seem to have been conducted on a scale even further beyond the
possibility of representation with a small cast and modest set of props.*

CHORUS

O for a Muse of fire, that would ascend
The brightest heaven of invention,
A planet for a stage, monsters to act
And Time Lords to behold the swelling scene!
Then should the Warlike Doctor, like himself,
Assume the port of Mars; and on the field
The Meanwhiles and the troops of Neverwere,
The Nightmare Child, the Horde of Travesties,
The Skaro Degradations and their kin;
The Time War's troops. But pardon, gentles all,
The flat unraised spirits that have dared
On this unworthy scaffold to bring forth
So great an object: can this cockpit hold
The plains of Gallifrey? or may we cram
Within this wooden O the bowships
That prevented not the fall of Arcady?
O, pardon! since a single Dalek may
Attest in little place a million;
And let us, ciphers to this great accompt,

On your imaginary forces work.
Suppose within the girdle of these walls
Are now confined two powers imperial
One great with saucers and trans-solar discs
The other fast with transduction's barrier.
Piece out our imperfections with your thoughts;
Conceive this paltry glass a Whitepoint Star
Fitted to scourge Creation with its light.
Think when we talk of Daleks, that you see them
Soar in squadrons through Kasterborous
For 'tis your thoughts that now must power our ships,
Lay waste our worlds; effect our jumps in time;
Spilling a tale as long as Rassilon's
Into an hour-glass: for the which supply,
Admit me Chorus to this history;
Who prologue-like your humble patience pray,
Gently to hear, kindly to judge, our play.

THE TRUE AND MOST EXCELLENT COMEDIE OF ROMEO AND JULIET

It is generally agreed that Shakespeare's play about the doomed love of Romeo and Juliet took as its main inspiration the poem The Tragical History of Romeus and Juliet *by Arthur Brook. The play, as reproduced in Quarto and Folio, closely follows the narrative of the poem, using the same character names and ending (as is well known) with the death of Romeo (when he discovers Juliet in a state of apparent death after taking a sleeping draught) and the death of Juliet (after she awakes and discovers Romeo has died).*

However, it appears that under pressure from James Burbage to 'make dark tragedie light' Shakespeare prepared a second version of the play ('the story as it did truly unfold, by misfortune unmarred') with a happy ending, turning the play (which is highly comic for its first three acts) into an out-and-out comedy…

This extract forms one of the longest pieces in the Shakespeare Notebooks.

ACT IV, SCENE III – JULIET'S BEDROOM

JULIET
>O look methinks I see my cousin's ghost
>Seeking out Romeo that did spit his body
>Upon a rapier's point. Stay, Tybalt, stay!
>Romeo, Romeo, Romeo, here's drink, I drink to thee.

>*She falls upon her bed within the curtains.*
>*Trumpet, wheezing, groaning.*

Enter Doctor, Rory and Amy.

DOCTOR
No, wait! You must not sip the sleeping draught!

RORY
Too late, she's out as cold as winter night
Her eyes respondeth not unto the light
Her heart, it beats so slow and faint to touch,
Her breath's so slight, no feather would it stir
And in this state would be mistook for death

AMY
Can not she be up-roused from her sleep?
A slap upon her cheek or touch of ice?
Or clap my hands over her face?

DOCTOR
No good
She has partook potion enough to stun
An elephant. You could a petard bang
And she would not stir from her slumber deep.

RORY
Then what are we to do? We dare not stay.
To be discovered with Julie's corpse.

AMY
No, it would not look good if we were found.
They'd have us put to death for murder sure

DOCTOR
You're right, and so we three must hence away.

Exeunt Doctor, Rory and Amy. Trumpet, wheezing, groaning.

ACT V, SCENE I – A STREET IN MANTUA

The Apothecary gives Romeo a vial of poison.

APOTHECARY
Put this in any liquid thing you will
And drink it off; and if you had the strength
Of twenty men it would dispatch you straight.

Romeo gives him gold.

ROMEO
There is thy gold, worse poison to men's souls.
Farewell, buy food, and get thyself in flesh.

Exit Romeo.

APOTHECARY
Die well! *[aside]* Think thou that mine's a murderer's trade?
But hark you this; those that seek out my wares
Are satisfied with the exchange. At least,
I've not so far receiv'd any complaints!

Trumpet, wheezing, groaning.
Enter Doctor, Rory and Amy.

DOCTOR
Stop! O, do not buy the cursed vial!

RORY
Too late, Doctor, we must have just missed him.

AMY
Romeo! Romeo! Where then is he, Romeo?

APOTHECARY
>Methinks I did hear word of his intent
>But find myself distract with my hunger
>>*The Doctor gives him gold.*

DOCTOR
>Here take this gold and buy thyself some lunch.

APOTHECARY
>I thank you sir. I heard him say that hence
>He would to the Capulet tomb go straight
>To lie with his dead love call'd Juliet.

RORY
>You mean he's gone back to Verona now?

AMY
>His love you see is merely feigning death!

APOTHECARY
>Oh no!

DOCTOR
>Come friends there is no time to lose!
>We must with haste to TARDIS quickly fly!

Exeunt Doctor, Rory and Amy. Trumpet, wheezing, groaning.

APOTHECARY
>'Tis not for me to care what those who buy
>My goods may do in folly or remorse.
>It hath for me been a rich day of trade
>I shall now dine and drink my health in wine!

Exit Apothecary.

ACT V, SCENE III – CAPULET TOMB IN THE VERONA CHURCHYARD

Romeo has discovered Juliet lying on an altar in the tomb.

ROMEO

    Arms, take your last embrace, and lips, O you

    The doors of breath seal with a righteous kiss

    A dateless bargain to engrossing death.

    Come, bitter conduct, come, unsavoury guide.

    Thou desperate pilot, now at once run on

    The dashing rocks thy seasick weary bark!

    Here's to my love.

He reaches for the vial of poison.
Doctor, Amy, Rory appear from behind the altar.

DOCTOR

    Romeo, stop! Don't drink the poison'd brew!

    For if thou dost thou shalt regret the deed

    As long as thou shalt live; which won't be long

But that is not the point. The point is this;
Thy Juliet is not dead yet; she lives!

ROMEO

I see no breath, her cheeks are pale, her lips
Are cold as stone. My love is dead, so taunt
Me not; I am resolv'd to die. But wait.
Who are you that dares violate the tomb
Of Capulet? And what is this blue box
That is not of this place?

AMY

We will explain
That later on.

RORY

Just put that vial down.
You heard the Doctor's words. Your Juliet
Just counterfeits death's signs. She slumbers deep
But will soon wake to find you here. And would
You wish she found you dead at her bed-side?
As consequence of feigned death? What would
She do in such a state of discontent?

ROMEO

I dare not think.

AMY

She would do something rash
Like take your dagger and do herself in.

DOCTOR

And would not that be a grave tragedy?

ROMEO

A tragedy forg'd of a grave misdeed,
Within a grave itself is grave indeed.

RORY

You're making jokes? At such a time as this?

AMY

That is those born of Italy for you!

ROMEO

The thought that Juliet might live dost seem
A hopeless hope pluck'd from a madman's dream.

DOCTOR

She soon will stir, if you would bring her round.

ROMEO

What must I do? My mind is all a whirl.

AMY

It's obvious. You have to kiss the girl!

Romeo kisses Juliet.

ROMEO

She lives! She breathes! Her eye-lids part! Her skin
Doth gain a rosy hue. Her hands are warm
Her fingers move. She shakes off death's black veil!

Juliet rises.

JULIET

My Romeo. 'Tis you! You found me then!
A kiss from thee lends me the breath of life.
It heats my blood. Please do so once again.

Romeo kisses Juliet.

ROMEO

But Juliet I dost not understand
Why you should play at death in this dark tomb
And risk your love to find you in a sleep
Of death? What were you thinking of, forsooth,
To hear that you were dead I poison bought
And was about to take it ere you woke.

JULIET

You did not get the letter that I sent?

ROMEO

What letter? I received none.

JULIET

The one
The Friar sent to Mantua for you?

ROMEO

I saw it not, I did not tarry there.

DOCTOR

It matters not. You're both here now and fate
Has been re-writ.

JULIET

Then we must flee this place
And start a life far from Verona's walls.

DOCTOR

Don't go just yet. You see you must first heal
The rift between the house of Montague
And Capulet.

ROMEO

And how shall that be done?
'Tis impossible.

AMY

> Not quite, you see, we have
> A cunning plan.

RORY

> What if you had both died
> Tonight and had in death been discover'd?

DOCTOR

> Such tragedy would show hate's consequence
> And teach them both to end their harsh discord
> And enmity. And so there will be peace
> In Verona at last.

JULIET

> But now we live
> That reconciliation is undone.

AMY

> Not necessarily. Because we have
> Another Romeo and Juliet!

> *Second Romeo and Juliet emerge from blue box.*

DOCTOR

> If you could step out of the way then they
> Will lie where you both would have lain in death.

> *Romeo and Juliet move as Second Romeo and Juliet lie
> down, Romeo on the altar, Juliet lying across him.*

ROMEO

> Who are these that like Proteus doth take
> The mirror'd semblance of my love and me,
> That walk and dost not speak and counterfeit
> The presentation of our dead likeness?

JULIET

> They cannot be of human flesh and blood,
> Our witch-craft summon'd twins! They dost not breathe!

RORY

> Fear not, the Romeo is but a clone
> Within Sontaran vat recently grown.

AMY

> And Juliet is not some fearful spectre
> She is in fact a borrow'd Teselecta.

ROMEO

> I do not understand, thy words are strange.
> But soft! I hear some noise. The watch approach!

Alarums.

DOCTOR

> Anon, we must inside our craft withdraw!

JULIET

> What craft?

RORY

> The TARDIS, it's the box of blue.

ROMEO

> But there will not be room enough for us!

AMY

> Oh you have no idea, young lover boy.
> Don't stand around, that means your girlfriend too.

JULIET

> I'm not his girlfriend, I'm his wife, thank you.

AMY

How old are you? Oh, never mind. Get in!

DOCTOR

For goodness' sake, could you stop arguing!

Doctor, Amy, Rory, Romeo and Juliet enter blue box.
Alarum. Watchmen enter and discover the bodies of the
Second Romeo and Juliet.

CHIEF WATCHMAN

Pitiful sight! He lies Romeo slain,
And Juliet bleeding, warm and newly dead.
Who here hath lain this two days buried.

Enter another Watchman with Friar Laurence
as their prisoner.

WATCHMAN

Here is a friar that trembles, sighs and weeps.
We took this mattock and this spade from him
As he was coming from the churchyard's side.

CHIEF WATCHMAN

A great suspicion! Stay the friar.

Enter Capulet, his wife and Juliet's Nurse.
They see the bodies.

CAPULET

O heavens! O wife, look how our daughter bleeds!
This dagger hath mista'en, for lo, his house
Is empty on the back of Montague,
And is mis-sheathed in my daughter's bosom.

NURSE

O me, this sight of death is a bell
That warns my old age to a sepulchre.

Enter Montague and his wife. They see the bodies.

MONTAGUE

O Romeo! What a sight is this,
To press before thy father to a grave?

CAPULET

O brother Montague, give me thy hand.
This is my daughter's jointure, for no more
Can I demand.

MONTAGUE

But can I give thee more,
For I will raise her statue in pure gold,
That whiles Verona by that name is known,
There shall no figure at such rate be set
As that of true and faithful Juliet.

CAPULET

As rich shall Romeo's by his lady's lie,
Poor sacrifices of our enmity.

*Montague and Capulet shake hands and hug,
their rift healed.*

CHIEF WATCHMAN

Bring forth the party of suspicion
And say at once what thou dost know in this.

Watchmen bring forward Friar Laurence.

FRIAR LAURENCE
I will be brief, for my short date of breath
Is not so long as is a tedious tale.
Romeo, there dead, was husband to that Juliet,
And she, there –

Doctor emerges from blue box.

DOCTOR
Hold your horses monky-boy!
Don't ring the undertaker, not just yet!
'Cos that's not Romeo and Juliet!

CAPULET
Who are you that dares skulk inside our tomb?

MONTAGUE
He must have slain our offspring! Seize him, guards!

DOCTOR
No one's been slain as thou will shortly see.
OK, it's time for you to come out now!

*Romeo and Juliet emerge from the blue box,
followed by Amy and Rory.*

FRIAR LAURENCE
Another Romeo and Juliet!

ROMEO
We are the true, those corpses are not us.

JULIET
They are but effigies in our likeness.

Capulet embraces Juliet.

CAPULET

If there be truth in sight, you are my daughter.

Montague embraces Romeo.

MONTAGUE

If there be truth in sight, you're Romeo.

ROMEO

I'll have no father if not he you be.

MONTAGUE

So you are wed! In thy example see
Our houses restor'd love and amity.

Romeo and Juliet join hands.

ROMEO

But there is still a blemish on our joy
The demise of Tybalt and Sir Paris
Both by my sword, much to my own remorse
I must from Verona be banished
I only ask that for my crimes my wife
Should not be punished.

DOCTOR

I'll stop you there!
Because I have a small surprise in store.

Paris emerges from the blue box with Rosaline.

PARIS

I am unmurdered as you can see,
So Romeo for that is not guilty.

ROMEO

But how? Methinks I did mistake you not
And ran you through with deathful wound just now.

PARIS

It was not I you fought and blood beget.

DOCTOR

That was in fact a Nestene duplicate.

PARIS

No longer wish I Juliet was mine,
Now that I have found love with Rosaline.

Paris and Rosaline join hands.

ROMEO

But for the death of Tybalt I repine.

DOCTOR

I think you'll find that he is also fine.

Tybalt emerges from the blue box.

FRIAR LAURENCE

And it seems meet that on this blessed night
I also should my nuptial troth-plight
My hope is that my suit is not adverse
When offered to my secret love, the Nurse.

NURSE

I did not have suspicion of your lust,
But will I take your hand in mine? I dost.

Friar Laurence and the Nurse join hands.

AMY

    But Doctor, how can Tybalt now live on?

DOCTOR

    'Tis not Tybalt, Amy, but a Zygon.
    I saved it from a burning as a witch
    Now it returns the favour owed.

RORY

    Neat switch.

DOCTOR

    And so the couples are all set to wed
    In fortune where mischance had once misled.
    Our work is done, 'tis time for us to leave
    Because no one our story will believe.
    'Tis said no tale could hope to overset
    The love of Romeo for Juliet,
    But never was there a more joyful story
    Than that of Amy Pond and her dear Rory.

Rory and Amy join hands. Exeunt.

Oh that this too too solid TARDIS would melt, thaw, and dissolve itself – adieu.

THE TEMPEST – A WORK IN PROGRESS

This extract appears to comprise Shakespeare's working notes for The Tempest. *However, the notes appear to be by two different writers – the different handwriting distinguished here by the use of italic text for the author generally acknowledged* not *to be Shakespeare.*

For obvious reasons, this extract more than any other has been cited by sceptics as anachronistic proof that the Notebooks are not genuine.

Shipwreck story – cash-in on Strachey's "Sea Venture" – check Middleton not already doing something.

~~The Shipwreck~~ ~~All At Sea~~ ~~The Storm~~ ~~The Strong Wind~~ ~~The Very Strong Wind~~ ~~The Gust~~ ~~The Extremely Strong Wind~~ ~~The Squall~~ The Tempest (good!) ~~The Very Strong Tempest~~

Also – Blackfriars have nets / lobster pots / sand dune backdrop from "Pericles" and "Winter's Tale". Note to self: Make sure they don't use real fish this time. Gets very smelly very quickly. Also waste of fish.

Love story – boy/girl – feuding fathers (been done?) – Kings or Dukes of somewhere in Italy (not Verona, not Venice, maybe Milan? Naples?)

Clown business – funny sailor? Shoemaker? (Why would a shoemaker be at sea?) Soldier? Funny butler? Working for Duke? Tailor? Cook? Potential double-act. Two funny butlers? ~~Priest?~~ Duke's Jester?

Political intrigue – feuding fathers are both Dukes of Milan – one has deposed the other (good!!!) deposed Duke exiled to island, usurper shipwrecked there, ends with deposed Duke slaughtering usurper! Revenge tragedy. No. Brief is for romantic comedy with masque section. Deposed Duke good guy, forgives usurper, boy and girl get married, everyone goes home. (But why did usurper exile deposed Duke and not kill him when he had the chance? Potential plot hole, need to think up good reason!)

Twist - put shipwreck at the beginning! Set whole thing on island. Show Johnson I can do Aristotelian unities. Action takes place in real-time! (Call it "Two Hours"? No)

Dukes could be brothers! Family angle. Also: explains why not keen to kill each other. Usurper Duke helped by King of Naples. (Check to make sure Naples is near Milan! No more geography cock-ups!)

Note to self: try to work in chess scene. Chess popular craze, always get sponsorship. Dukes play chess? Chess-themed song and dance? ~~Musical about chess?~~ Boy and girl play chess?

On island, usurper Duke tries to kill King of Naples, deposed Duke stops him – clever – nice irony.

Dukes' wives – dead or at home in Naples. What about crew of shipwrecked ship? Bit grim if all drowned. Think comedy, keep it light! Just lost on island? Only Duke and King swept overboard?

Ship undamaged so they can all go home at the end? Bit implausible. Also: big coincidence that usurper Duke on same island as deposed Duke. (But got away with it in "Comedy" and "Twelfth"!)

Deposed Duke used to be wealthy. Needs Rich-sounding name. ~~Wealthio? Magnifico?~~ Prosperous. Prospero.

Usurper Duke – ~~Usurpio? Mutinio~~ Antonio (placeholder only, have used that name four times before!)

Needs to show off Blackfriars theatre new special effects. Off-stage sound-effects and music, trap-doors, surprise water cannon.

Comedy double-act get drunk (on wine from shipwreck) – funny dressing-up scene – enormous trousers always hilarious.

Missing something. Pirates? ~~Primitives worshipping giant squid~~ (impractical – could do monkey)? Cannibals? Cannibals worshiping monkey? Worshipping bear? (Note to self: use man in bear suit. Do not use real bear after last time.) ~~Polar bear? Sinister black smoke? Mysterious hatch.~~

Feels like I'm repeating myself. Maybe I'm getting too old for this game, running out of ideas, a bit played-out. Get young Johnny Fletcher in to co-write? <u>Not George Wilkins</u> again.

Or just quit while I'm ahead, retire to the country? Definitely needs another element, not enough material for five acts. All a bit ordinary, going-through-the-motions.

You're right, it needs the wow factor.

What's this? I didn't write this.

No, I did.

And who are you, to tamper with my notes while I am asleep?

The Doctor. Sorry can't be with you in person, just missed you by a few months, and I can't hang around. Small dinosaur situation. Well, it was small to begin with. They grow. So I popped in to see Jack the paper merchant – lovely fella, Jack – and I'm writing on the sheet of paper you'll buy in two months' time. I've got him to put it aside for you specially. Or I will, Will.

But your words weren't there yesterday. And how do you know <u>what</u> I will write on the paper, for you to reply to it?

Because I also have a copy of this manuscript from the future. Picked it up while I was doing some proofreading on the first Folio. Your spelling, I don't want to criticise, but get a dictionary! And once I'm done I'll probably post it anonymously to the British Library or something. Give the experts something to argue about.

What first folio? You cannot mean somebody intends to publish my works without my permission?

Don't worry about it. Point is, I have a copy of this piece of paper from the future, so as I write on this one, the words appear on the one __from__ the future. Bit timey-wimey, best you don't think about it too much.

What if I were to destroy this paper now, tear it into pieces and burn it in the fireplace?

Don't. Please, don't. Fabric of time is in a bad enough state as it is. And besides, you can't throw it away, you have the workings of a great play here. Given a bit of help.

And that is why you have chosen to intercede in the creative process?

Yep. Go to Jack's, I've left a parcel with him. Should all be fairly self-explanatory.

I have done so. It contains a silver disc and a folding black volume. It does not explain itself.

It's a battery-powered Blu-ray player, all charged up. You just press the button, put the disc in and it'll play. Hope you enjoy the movie.

"Movie"? And you take issue with my spelling? But I have watched this strange moving presentation, "Forbidden Planet". It is a little far-fetched and some of the concepts are unfamiliar, but the central premise is intriguing. Indeed, it bears some similarities to what I intended to do with this play.

That's the point. It is. It's inspired by the play you're about to write. So feel free to raid it for ideas because, in a way, they're your ideas, so it's not stealing.

You say I should take inspiration from a "movie" based upon my work? So that the makers of the "movie" can then draw inspiration from my play? Without paying me?

Yeah, but by then you're out of copyright, so there's nothing you can do about it. Thing is, you have to write "The Tempest", or they won't be able to make the film, and, well, fabric of time and all that. And anyway, you want to go out with a bang, don't you? Good luck. Gotta go now. Donna's getting impatient and we have that whole dinosaur thing to deal with.

The Tempest – New thoughts

Keep the island, keep the shipwreck, keep the surprise water cannon.

Prospero is not just Duke. He is also a ~~magician~~ wizard who studies books of magic. Took them with him to island? Or developed magic powers on island? Both?

Prospero gets powers from ~~vast ancient alien underground machine magic tree~~ magic ~~stick branch~~ staff!

He causes storm at beginning to <u>bring</u> usurper Duke to island (good!). Removes coincidence. Also magic shipwreck to explain why only usurper Duke, King of Naples, Duke's son, courtiers and funny servants come to shore. Rest of crew and boat on other side of island, everybody asleep.

Prospero has metal man to carry out orders. "Robot" as in the "movie"? Too much like man in suit of armour, audience will never buy it. Fairy? Familiar spirit? Needs windy name? ~~Wafty?~~ ~~Gust?~~ ~~Aireo?~~ Arial? Ariel perhaps.

(Check to see if Puck costume still in stores)

Love story – Girl is Prospero's daughter, exiled to island with him, has never seen any other men, love at first sight. Prospero uses ~~Puck~~ Ariel to put boy through the mill a bit to check he's worthy of his daughter. Girl is overcome when she sees other arrivals, "oh ~~lovely~~ brave new world that has such ~~boys~~ ~~men~~ people in it". Work in chess game.

Prospero uses ~~Puck~~ Ariel to sort out court intrigue, saves King of Naples life. Using "force field" to make weapons freeze in air etc.

Invisible id monster attacking "force field" – not sure even Blackfriars up to it. But invisible idea good. ~~Puck~~ Ariel is invisible (but we can see him). Do people-being-pinched-by-invisible-attacker gag (been ages since we did it in "Midsummer's Night Dream", no one will remember). Also fairy can ~~shove clowns into rivers~~ lead clowns into swamp.

Can also show off acoustics of Blackfriars, Ariel singing from behind audience, music coming from hidden musician's gallery. Masque section – Prospero summons up invisible band and more dancing fairies. Appear and vanish using trapdoors! Insubstantial pageant. Sounds and sweet airs.

Still needs "baddie". Monster is ancient alien that built machine? Or primitive? Needs cannibally name. Think anagrams. Whole tribe or only one? Check budget.

Instead of ancient alien machine, a witch was exiled to island (good!). Canibal was her servant / son. She created Ariel. Then died, leaving Canibal alone on island. (What happened to Ariel?

Buried in fathoms of the Earth? Or in sea five fathoms deep? <u>Or trapped in magic tree</u>?)

Set free by Prospero, hence debt of gratitude. At end, Prospero sets Ariel free, gives up magic ~~stick~~ staff and books. Time to return to real world, dream is over etc.

Pair up Canibal with clown double-act. Get him drunk on wine, dress up in hilarious trousers (another of Ariel's magic pranks). Strange bedfellows!

Big show-down between Prospero and witch at end. Duelling magic staffs. Lots of explosions, smoke, trapdoor business. (Note: <u>check we still have witch costume in store, or we may have to keep witch off-stage.</u>)

Name for witch? If this is my last play, might as well use that name suggested by Doctor as little thank you. Call her <u>Sycorax</u>!

So, three plot lines, court intrigue, love story, comedy business with clowns / Canibal. All sorted out by Prospero and Ariel. Too easy? Or claim it is deliberate "deux ex machina" to impress Johnson? Is there a gag in "Dukes ex machina"?

Nearly there, rest of story will write itself. Should be a real crowd-pleaser, short, sweet and simple! Just need to think of character names for clowns, etc.

Ah, might be able to help you with that. Here's a list of the moons of Uranus – Miranda, Ferdinand –

What is Uranus? That's not a word I would ever come up with, it sounds disturbingly like –

Planet, ice giant, seventh one out, won't be discovered for another eighty years so keep it to yourself. Oh, and one last thing. Don't forget to put this sheet in with your 'foul papers' so I'll come across it when proofreading the Folio, or, you know, fabric of time.

I will not forget, Doctor. And thank you for your help. I trust the fabric of time is secure. But I still need a name for the King of Naples. A name I have never used before.

Oh, I'm sure you'll think of something. Allons-y, Alonso!

ALONSO! Yes, that's it! Alonso! Alonso! <u>ALONSO</u>!

Now could I drink hot chocolate.
Through my special straw that makes it fizzy.

EXIT, BY ANOTHER MEANS

This alternative draft of a section of Act III Scene iii of The Winter's Tale *includes a very different exit for Antigonus after he leaves the baby Perdita to fend for herself. Famously, as published in 1623, the stage direction is: Exit, pursued by a bear. The manner in which Antigonus leaves in this version is rather different, but again it is clear that he has been removed from the action of the play and will not return.*

It is amusing to think that Shakespeare may have replaced this version with 'Exit, pursued by a bear' as he thought the audience would find that more plausible.

There lie, and there thy character; there these

Laying down a bundle

Which may, if fortune please, both breed thee, pretty,
And still rest thine. The storm begins. Poor wretch,

A wind blows up from nowhere, unearthly

That for thy mother's fault art thus expos'd
To loss and what may follow! Weep I cannot,
But my heart bleeds; and most accurs'd am I
To be by oath enjoin'd to this. Farewell!
For I must haste return to Gallifrey
Too long have I on Earthly shores been homed
'Tis time to leave. A savage clamour!

*The strangest sound akin to great wheezing and groaning
that ever did assail the ear. A blue box doth appear.*

'Tis the strangest sound to assail the ear
And now my blue box doth appear.
Well may I get aboard! This is the chase;
I am gone for ever.

Exit, into the blue box, which then vanishes amid great clamour

A TARDIS, a TARDIS,
my kingdom for a TARDIS...

THE WINTER'S TALE

Another draft extract from Act V Scene iii of The Winter's Tale – *this time part of the scene in the Chapel at Paulina's house that forms the climax of the play, as Paulina reveals what purports to be a statue of the dead Hermione. Of course, the audience knows that the statue is actually Hermione herself, who isn't dead at all…*

LEONTES
    O Paulina,
    We honour you with trouble; but we came
    To see the statue of our queen. Your gallery
    Have we pass'd through, not without much content
    In many singularities; but we saw not
    That which my daughter came to look upon,
    The statue of her mother.

PAULINA
    As she liv'd peerless,
    So her dead likeness, I do well believe,
    Excels whatever yet you look'd upon
    Or hand of man hath done; therefore I keep it
    Lonely, apart. But here it is. Prepare
    To see the life as lively mock'd as ever
    Still sleep mock'd death. Behold; and say 'tis well.

      PAULINA draws the curtain, and discovers
      HERMIONE standing like a statue

    I like your silence; it the more shows off
    Your wonder; but yet speak. First, you, my liege.
    Comes it not something near?

Enter a strange PHYSICIAN, wearing a hat of Afric.

PHYSICIAN
    Oh, good job I got here in time.
    You are as bad as Julius Grayle, you know,
    In keeping cutain'd statues hid from view.
    Think I wouldn't find you, did you – eh?
    Aye, well I've got news for you, Sunshine.
    The rest of you, keep looking on the scene
    Beware the statue that appears to be
    Cold stone, unmoving, safe and fancy free.
    I warn you not to blink or look away

Your lives depend on what I tell you now
Don't blink, don't blink – whate'er you do don't blink.

LEONTES
Methinks the fellow is sore vexed at heart.

PHYSICIAN
I tell you, vexed is but the start of it…
Hang on. Am I in the right place? This is 1904, isn't it, or has my watch stopped? Again? Just a minute – you're not even a real statue. What's going on here? Some jolly japes is it? Party games? Pin the tail on the… No, sorry, I'm spoiling things aren't I? This was meant to be a big surprise, climactic reveal and everything. And my iambic pentameters just went right out the window. Sorry everyone. Especially you, Miss Statue. I should… I'll just be going now, if that's all right with everyone? Um, bye then.

Exit the PHYSICAN.

*To reverse or not to reverse
the polarity of the neutron flow?*

ANTONY AND CLEOPATRA

This early draft extract from Antony and Cleopatra *appears without explanation in the Notebooks. In this scene, Cleopatra, Queen of Egypt, defeated in battle by Caesar and mourning for Antony, has retreated to her monument, where she plots to kill herself.*

ACT V, SCENE II

…Enter Clown with basket

CLEOPATRA
Hast thou the pretty worm of Nilus there
That kills and pains not?

CLOWN
Truly, I have him; but I would not touch him, for his biting
is immortal. This is most falliable, the worm's an odd worm.

CLEOPATRA
I thank you. Farewell.

CLOWN
Indeed, there is no goodness in the worm.

CLEOPATRA
Yes, yes, get thee gone. Farewell.

CLOWN
One word more, my queen. Put up thy hand.

CLEOPATRA
For sooth! Would you tell me my fortune?
I hazard I can do that well myself.

CLOWN

Be not so sure, good lady. Fortune turns her wheel. A clown
shall be a king and a king shall be a clown. The worm will do
his kind. Behold.

Uncovers basket.

CLEOPATRA

Fair snake, are you ripe fortune's worm, indeed?

CLOWN

The Mara offers you its hand. I follow the snake and the
snake follows me. Such is fate. This worm's an ouroboros that
eats its own tail and is never full. Fortune turns her wheel.
Civilisations rise and civilisations fall. And the worm eats all.

CLEOPATRA

What a poetic fool art thou!

CLOWN

    Yesterday a soothsayer. Today a fool.
    I am what the worm bids me. Look at me.

CLEOPATRA

    A queen look at a clown? Go to.

CLOWN

    Look at me. I'm not trying to harm you.
    Look at me.

CLEOPATRA

    Tis a basilisk glare!
    I like not that. There's hunger in thy stare.

CLOWN

    Your path's as forked as is my tongue.
    One way leads down, the other up the hill.
    Take my hand and we shall rise together.
    For my snake has wings that turn the wheel.

CLEOPATRA

    Magick snake indeed. You do seem taller.

CLOWN

    My snake gives me power. I serve it well.
    It serves me better. We are well match'd.

CLEOPATRA

    I am in thrall to no one.

CLOWN

    No? Good queen,
    You've saddled to countless Emperors.
    Why not ride with a snake to generation?

CLEOPATRA

> Couple with a serpent? I am no fool.
> I do not fear the asp's bite. I fear its promise.

CLOWN

> Look at me. Take my hand. And the snake shall
> Take thee whole. All is not lost. The wheel turns.
> You shall be right Egypt's queen again.

CLEOPATRA

> Royal Egypt by a gypsy gyped!
> I have won enough fortunes to not count
> The losing of them. You cannot restore
> What I most wish.

CLOWN

> Your kingdom? 'Tis no matter.

CLEOPATRA

> My Antony.

CLOWN

> In that you are too late.

CLEOPATRA

> Then what is a kingdom without a king?
> Antony's my horizon and my ruler.
> Cleopatra with no Antony is
> No Cleopatra. I would not be ruled
> By else. No snake can charm his place.

CLOWN

> Take my hand, fair queen. It will not bite.
> Shake hands with a snake? Ah that's less certain.
> A woman is a dish for the gods.
> Reach into the basket and find time's jaw.

CLEOPATRA

Ah, get thee gone. Farewell.

CLOWN

Yes, forsooth. I wish you joy o'th' worm.

Exit

CLEOPATRA

Give me my robe. Put on my crown. I have
Immortal longings in me.

That which we call a Rose
by any other name would still be Tyler.

TROILUS AND CRESSIDA

This is a peculiar variant version of Shakespeare's play about the Trojan War – a conflict that began when Paris, son of Priam, King of Troy, persuaded the Spartan Queen, Helen to leave her husband, Menelaus, and return with him to his home city. Troy was besieged by the Greeks for almost a decade – until the deadlock was broken by a clever trick involving a wooden horse.

Here, the Prologue from the original version of the play seems to have some trouble keeping to his place in the drama – and also claims to be the author of the successful Greek plan. Some scholars point to this as more anachronistic evidence that the Notebooks are fakes.

ACT I

PROLOGUE

In Troy, there lies the scene. I'm sure you know
The tale. Perhaps you've read it in a book,
Or seen it on your television set.
Perhaps a classics master, with a cane
Tapped out Homeric rhythm on your desk?
When Susan went to that ridiculous school
She carried home some volume from Miss Wright
That detailed what was known about this war.
(She sat upon her bed and rolled her eyes.)
We had a copy, too, of the long work
That Chaucer made of *Troilus and Criseyde*
Bought from a barrow on the Goldhawk Road.
And I recall a dismal afternoon
When I took refuge in the ABC
And saw the story screened in Cinemascope
With stuck-up British actors flouncing round

In dreadful togas all in pastel shades.
(Whatever happened to that Maxwell Reed?)
The maker Robert Henrysoun, I think,
Wrote the first version that I came across.
I said to him, "Now sir, does this poor girl
In *Testament of Cresseid* have to die
In such a nasty way?" I read it out.
(Excuse the change in metre as I quote.)
"My cleir voice, and courtlie carrolling,
Quhair i was wont with ladyis for to sing.
Is rawk as ruik, full hiddeous hoir and hace
My plesand port all vtheris precelling."
And if you have forgot your Middle Scots
He's writing of the scars of leprosy
Visited upon Cresseid as punishment
For faults, it seems to me, were not her own.
(I told him this was harsh, but he was set.)
So how did Shakespeare tell the story? Hmmm?
He started with a Prologue, as I do.
Uneasily, like this: "From isles of Greece
The princes orgulous, their high blood chafed,
Have to the port of Athens sent their ships,
Fraught with the ministers and instruments
Of cruel war: sixty and nine, that wore
Their crownets regal, from the Athenian bay
Put forth toward Phrygia; and their vow is made
To ransack Troy, within whose strong immures
The ravish'd Helen, Menelaus' queen,
With wanton Paris sleeps; and that's the quarrel."
This is supposed to be an epic play.
Old Homer in the *Iliad* produced
A list of ships of somewhat greater length:
One thousand, one hundred and eighty-six.
And "quarrel" seems a feeble choice of word

When Grecian heroes are protagonists;
Or so I thought, before the morning came
When my own ship, the TARDIS, made landfall
Upon the windblown plains of antique Troy.
And then I saw the truth of it myself.
This is how good Will Shakespeare sketched it out
In his fair play: "To Tenedos they come;
And the deep-drawing barks do there disgorge
Their warlike fraughtage: now on Dardan plains
The fresh and yet unbruised Greeks do pitch
Their brave pavilions: Priam's six-gated city,
Dardan, and Tymbria, Helias, Chetas, Troien,
And Antenorides, with massy staples
And corresponsive and fulfilling bolts,
Sperr up the sons of Troy." What do you think?
The list of names seems less than necessary:
You won't hear some of these again tonight.
But note that "fresh and yet unbruised" bit.
The Greeks are not quite heroes by this light.
It shows the playwright is a sharp young chap.
He saw the vein of the preposterous
That ran through this old saga and perceived
That it was starved of true nobility.
Adultery, a wrangle, jealous gods,
Two nations welded into leaden siege:
These were the real components of the tale.
I took the Prologue's part, but if you will
I'll not down to the tavern just quite yet.
Forgive me if I loiter in the wings
And act a kind of Chorus to our play.
Imagine me on foot across the plains,
But keeping clear from loftier terrain.
I'm not a mountain goat and I prefer
Walking to it any day. (And I hate

Climbing!) But I fall from my text, I fear.
Let me revert to what I have writ here:
"Now expectation, tickling skittish spirits,
On one and other side, Trojan and Greek,
Sets all on hazard: and hither am I come
A prologue arm'd, but not in confidence
Of author's pen or actor's voice, but suited
In like conditions as our argument,
To tell you, fair beholders, that our play
Leaps o'er the vaunt and firstlings of those broils,
Beginning in the middle, starting thence away
To what may be digested in a play.
Like or find fault; do as your pleasures are:
Now good or bad, 'tis but the chance of war."

A further fragment from later in the play also appears in the Notebooks:

ACT II, SCENE II – TROY. A ROOM IN PRIAM'S PALACE

PROLOGUE appears within an urn.

PROLOGUE
    I beg you, do not titter, if you please.
    Your Prologue is within the Trojan walls.
    And he is desirous not to be seen
    Particularly by Priam, King of Troy –
    Unmanly creature that I can't abide! –
    A churl who dines on peacock's breast and fills
    The luckless air with boring anecdotes.
    Soft, here he comes, in noble company.
    Let me remain cabin'd within this urn.

Enter PRIAM, HECTOR, TROILUS, PARIS,
and HELENUS

PRIAM

 After so many hours, lives, speeches spent,
 Thus once again says Nestor from the Greeks:
 "Deliver Helen, and all damage else –
 As honour, loss of time, travail, expense,
 Wounds, friends, and what else dear that is consumed
 In hot digestion of this cormorant war –
 Shall be struck off." Hector, what say you to't?

PROLOGUE (Aside)

 He would do well to heed these generous words
 If not, before this fateful week is out,
 He'll look some other gift horse in the mouth.
 And now speaks Hector, champion of Troy.

HECTOR

 Though no man lesser fears the Greeks than I
 As far as toucheth my particular,
 Yet, dread Priam,
 There is no lady of more softer bowels,
 More spongy to suck in the sense of fear,
 More ready to cry out "Who knows what follows?"
 Than Hector is.

PROLOGUE (Aside)

 Soft bowels? Is this good?
 He gallops to his point.

HECTOR

 Let Helen go:
 Since the first sword was drawn about this question,
 Every tithe soul, 'mongst many thousand dismes,
 Hath been as dear as Helen; I mean, of ours:

If we have lost so many tenths of ours,
To guard a thing not ours nor worth to us,
Had it our name, the value of one ten,
What merits in that reason which denies
The yielding of her up?

PROLOGUE (Aside)
Nay, and thrice nay!
Hector has his contention by the reins.
Yield up Helen of Sparta to the Greeks?
Perhaps the horse, now stabled in the trees.
Will never now be loosed upon the plains.
I must confess a feeling of dismay:
The blueprint of this equine strategy
Flowed from my pen within a Grecian tent –
Once I had been obliged to set aside
A better strategy founded upon
Catapulting the Greeks over the wall.
So now this fellow Troilus speaks.

TROILUS
Fie, fie!
Weigh you the worth and honour of a king
So great as our dread father in a scale
Of common ounces? Will you with counters sum
The past proportion of his infinite?
And buckle in a waist most fathomless
With spans and inches so diminutive
As fears and reasons? Fie, for godly shame!

HECTOR
Brother, she is not worth what she doth cost
The holding.

CASSANDRA

[Within] Cry, Trojans, cry!

PRIAM

What noise? What shriek is this?

PROLOGUE (Aside)

Now this I know, I met her by the walls.
Cassandra is she called. A lady who
Can gaze into the future's murky glass.
Or so she says. (Or shouts, I ought to say.)

TROILUS

'Tis our mad sister, I do know her voice.

CASSANDRA

[Within] Cry, Trojans!

HECTOR

It is Cassandra.

Enter CASSANDRA, raving

CASSANDRA

Cry, Trojans, cry! Lend me ten thousand eyes,
And I will fill them with prophetic tears.

HECTOR

Peace, sister, peace!

CASSANDRA

I heard a sound like thunder from the gods,
And from Troy's walls I gazed upon the plain.
There I beheld an unexpected sight:
A wooden tent, surmounted by a lamp.
And when its light ceased streaming o'er the sand
A strange old man came tripping from within.

Last night this man came wandering through my dreams
With hair like snow from Mount Olympus' peak,
And brow as stern as Zeus when angered.
Quoth he, "I have been drawn down from the stars,
– The sphere in which I seek my destiny –
And all the world about me now displayed
Is as a scene created for the stage.
Can this be Ulysses? Agamemnon?
The heroes of a thousand picture books?
I marvel at invention's power to draw
Such noble art from poor material."
And off he went, pocking the plains of Troy
With the end of his little wooden staff.
Wherefore do I harp on this old man's words?
I sense he brings destruction to us all,
In concert with the theft that sparked this war.
Troy must not be, nor goodly Ilion stand;
Our firebrand brother, Paris, burns us all.
Cry, Trojans, cry! A Helen and a woe:
Cry, cry! Troy burns, or else let Helen go.

Exit

PROLOGUE (Aside)
A narrow squeak! I thought she meant to lift
The lid of my disguise, exposing me
Like a stuffed partridge 'neath a silver cloche.
'Tis time for me to quit King Priam's halls,
Before the siege comes to a fiery end
And Agamemnon's flames have the effect
Of boiling me inside this goodly pot.
The lady whom her brothers said were mad,
She speaks the hardest truth; old Troy will burn
And nobody will ask: what's a Greek urn?

PERICLES

This sequence from the play Pericles *again appears with no explanatory notes within the Notebook. The introduction of the character 'Romana' has given rise to various theories as to why the sequence was reworked for the final version of the play.*

SCENE V – MYTILENE. A HOUSE OF ILL-REPUTE

Enter, from the house, two Gentlemen

FIRST GENTLEMAN
  Did you ever hear the like?

SECOND GENTLEMAN
  No, no. Come, I am for no more bawdy-houses: shall's go hear the vestals sing?

FIRST GENTLEMAN
  Aye. I'll do any thing now that is virtuous.

Exeunt

SCENE VI – THE SAME. A ROOM IN THE HOUSE

Enter PANDAR and BAWD

PANDAR
  Well, I had rather than twice the worth of her had she ne'er come here.

BAWD
  Fie upon her! She would make a puritan of the devil!

PANDAR

'Faith, is there no way to be rid of her?

BAWD

Soft. Here comes the Lord Lysimachus disguised.

Enter LYSIMACHUS

LYSIMACHUS

How now! What wholesome iniquity have you that a man may deal withal?

BAWD

We have here one, sir, if she would – but there never came her like in Mytilene.

LYSIMACHUS

Well, call forth, call forth.

BAWD

Never plucked yet, I can assure you.

Re-enter PANDAR with ROMANA

PANDAR

Is she not a fair creature?

LYSIMACHUS

'Faith, she would serve after a long voyage at sea.
Well, there's for you: leave us.

BAWD

Come, we will leave his honour and her honour together. Go thy ways.

Exeunt BAWD and PANDAR

LYSIMACHUS
  Now, pretty one, how long have you been at this trade?

ROMANA
  What trade, sir?

LYSIMACHUS
  Why, I cannot name't.

ROMANA
  I should think not. I hear say you are of
  Honourable parts. Here's no part of honour.
  I hear you are this country's governor.
  How you should rule it that can'st not rule thyself.

LYSIMACHUS
  How's this?

ROMANA
  Your mask is but a masque, trust me.
  A stranger here, most ungentle fortune
  Has placed me in this sty. I'm used to it,
  Not this place, but my sad condition.
  I travel on time's seas, and often am
  Washed up roughly on ungentle shores and
  Trapped in vile dungeons. But this is a first.

LYSIMACHUS
  You claim you are a gentlewoman?

ROMANA
  Lady Romanadvoratrelundar.
  And you rule Mytilene?

LYSIMACHUS
  I do.

ROMANA
Then check thy privilege and rule better.
Dos't hold thee in thy prisons a Doctor?

LYSIMACHUS
Need you physic?

ROMANA
Not so much as he will.
He is my companion, a wand'ring fool.

LYSIMACHUS
Oh, him.

ROMANA
I thought as much. Fetch him hither.

LYSIMACHUS
Shall I take commands of you?

ROMANA
I think so.
Someone needs to tell thee how to rule.
We came here by mischance but it can serve
Its turn. We search for a precious jewel.

LYSIMACHUS
I had hoped to take that from you.

ROMANA
Stop that.
Clearly it's a dead end. Fetch the Doctor,
Let me out, nail these bawds and mend thy ways.
Your cloak needs a press and there's egg upon thy cuff.
In the market lies a girl, Marina.
She leads a simple life, but her head's screwed on.
And she really is a princess. She'll do.

LYSIMACHUS
A princess? 'Tis said I'm quite the catch.

ROMANA
You'd be lucky to have her. She's got style.

LYSIMACHUS
How came she to land on fair Mytilene?

ROMANA
Oh, bad luck. Same as everyone else.

LYSIMACHUS
Harsh.

ROMANA
But fair.

LYSIMACHUS
You've a good point there, I fear.

ROMANA
Woo the girl, and then we'll see. Your state
Is in a state, but it's not up to me.
Mytilene needs Marina, as do you.
So chop chop. Fetch her forth, get the Doctor
Then come back here. I haven't got all day.

LYSIMACHUS
Will there be anything else?

ROMANA
Oh, bring gold.
They seem to like that here. The food's not bad.
I owe them that at least.

LYSIMACHUS
I'll pass that on.

ROMANA

Do get a move on, there's a dear. This dress
Is thin and there's a nip in the night air.

LYSIMACHUS

You have spoke most well; I never dream'd thou couldst.
Had I brought hither a corrupted mind,
Thy speech had alter'd it. Hold, here's gold for thee:
Persever in that clear way thou goest,
And the gods strengthen thee!

ROMANA

Please hurry! Spit-spot!

LYSIMACHUS

Fare thee well. Thou art a piece of noble virtue
Hold, here's more gold for thee.
A curse upon him, die he like a thief,
That robs thee of thy goodness! If thou dost
Hear from me, it shall be for thy good.

Enter PANDAR

PANDAR

I beseech your honour, one piece for me.

LYSIMACHUS

Avaunt, thou damned door-keeper! Away!

Exit

PANDAR

How's this? Another one? Oh, Romana,
We must take another course with you
We'll have no more gentlemen driven away.

Re-enter Bawd

BAWD
How now! what's the matter?

PANDAR
Worse and worse, mistress; she has here spoken
holy words to the Lord Lysimachus.

BAWD
O abominable! Marry, hang her up for ever!

PANDAR
She sent him away as cold as a snowball;
Saying his prayers too.

BAWD
Would she had never come within my doors!

ROMANA
I am still here, you know. Hello!

PANDAR
What now?

ROMANA
Just that rescue will be here in an hour.
I wonder if I could have some more stew?
It's rather good. Why not close up thy stew
And find something else to do?

PANDAR
What, prithee?

ROMANA
Have you heard of a good restaurant?
With linen on the tables, a menu
Printed, a fire, and lovely cosy chairs.

It's easy work and I think you'll clean up.
Let me show you how to fold a napkin…

BAWD

Marry, hang you! She's born to undo us.
Marry, come up, my dish of chastity
I'll serve thee up with rosemary and bays!

ROMANA

The day's been long. That had better be a yes.

Exeunt

*We are such stuff as dreams are made on;
and the Mara rounds on us in sleep.*

CORIOLANUS

The extant version of Coriolanus *is an account of the Roman military hero, Coriolanus, who is raised, unwillingly, to the highest office – then loses his political power and goes into exile in the country of his former enemies, the Volsces.*

This previously unknown version from the Notebooks is set in another territory entirely, referred to in the text as Tara, a possible allusion to the kingdom of ancient Ireland.

ACT II, SCENE II

Enter Thorvald, nobleman of Tara; the Archimandrite, an official; Kurster, captain of the guard; Till, a retainer.

THORVALD
    Proceed, Archimandrite.

ARCHIMANDRITE
    I shall lack voice: the deeds of Coriolanus
    Should not be utter'd feebly. It is held
    That valour is the chiefest virtue, and
    Most dignifies the haver: if it be,
    The man I speak of cannot in the world
    Be singly counterpoised. At sixteen years,
    When Grendel sought to seize the Taran throne
    By keeping good Prince Reynart in his cell
    And Princess Strella too, endungeon'd there,
    He swam the moat to bring a message through,
    Conveyed from royalist conspirators
    Who plotted at their lodge among the trees.
    He braved the guards of Gracht; he dodged their bolts
    And the amours of Madame Lamia,

The secret consort of the wicked count.
He faced the perils of the Taran night
And slew a knot of Wood Beasts with his blade.
And when the great alarums fiercely broke
On hated Grendel's fateful nuptial night
He loosed his crackling sword out of its sheath
And battled bravely on. In that day's feats,
When he might act the woman in the scene,
He proved best man i' the field, and for his meed
Was brow-bound with the oak. His pupil age
Man-enter'd thus, he waxed like a sea,
And in the brunt of seventeen battles since
He lurch'd all swords of the garland.
In these dark times, now Reynart's kingdom finds
It has new enemies beyond the trees,
He lent our troops a spark of his great fire
And by his rare example made the coward
Turn terror into sport: as weeds before
A vessel under sail, so men obey'd
And fell below his stem. As if one mass,
They surged across the bloodied plains
Of Thorvald and Mortgarde, and thence rode on
To Freya, where they first caught sight of the
Dread foe who had laid waste these lands. We have
Not name for them, but those who saw their flight
Spoke of steel bodies and a burning eye;
Curved ribs of metal, and a stubby rod
That shot a burning bolt of hot brimstone
Through plated ranks of rosy Taran youth.
One man, Coriolanus, was not moved.
He rode on, bravely, thus. His sword, death's stamp,
Where it did mark, it took; from face to foot
He was a thing of blood, whose every motion
Was timed with dying cries; the gurgles of

The hidden occupants of these steel shells.
Still on he rode: unto the city that
Our metal foes had planted 'pon the plain.
A gilded plate suspended there on stilts,
Massy; a league from edge to edge, they said.
And from its glassy portals could be seen,
More of our foes, bright lamps upon their skulls,
Gazing dispassionate upon the sight.
Brave captain ours; alone he entered
The mortal gate of the city, which he painted
With shunless destiny; aidless came off,
And with a strange sword roaring in his fist
Ran reeking o'er metallic dead, as if
'Twere a perpetual spoil: and till we call'd
Both field and city ours, he never stood
To ease his breast with panting.

TILL

Worthy man!

KURSTER

He cannot but with measure fit the honours
Which we devise him.

THORVALD

Our spoils he kick'd at,
The ground was thick with riches from the stars
Minerals and devices crystalline,
Lay scattered on the muddy Taran ground.
He look'd upon things precious as they were
The common muck of the world: he covets less
Than misery itself would give; rewards
His deeds with doing them, and is content
To spend the time to end it.

KURSTER

> He's right noble:
> Let him be call'd for.

TILL

> Call Coriolanus.

KURSTER

> He doth appear.

> *Enter CORIOLANUS, attended by ZADEK*
> *and FARRAH, two swordsmen.*

ARCHIMANDRITE

> The senate, Coriolanus, are well pleased
> To crown thee Tara's king.

CORIOLANUS

> I do owe them still
> My life and services.

ARCHIMANDRITE

> It then remains
> That you do speak to the people.

CORIOLANUS

> I do beseech you,
> Let me o'erleap that custom, for I cannot –
> I cannot - not - not -

FARRAH (Aside)

> Methinks I heard a spark. The hero has
> A wire crossed within its circuitry.
> What should we do? Old Zadek, quickly, speak.

ZADEK

> Our hero is exhausted by his feats
> He is a man of action, not of words.
> Forgive his silence; call it eloquence.

KURSTER

> Sir, the people
> Must have their voices; neither will they bate
> One jot of ceremony.

ARCHIMANDRITE

> Pray you, go fit you to the custom and
> Take to you, as your predecessors have,
> Your honour with your form.

CORIOLANUS

> It is apart
> That I shall blush in acting, and might well
> Be taken from the people.

ZADEK

> Mark you that? How modestly he speaks.

FARRAH

> How fit he is to lead the Taran state!

CORIOLANUS

> To brag unto them, thus I did, and thus;
> Show them the unaching scars which I should hide,
> As if I had received them for the hire
> Of their breath only!

FARRAH (Aside)

> His voice returns. What strategy, Zadek?

ZADEK (Aside)

The Doctor, 'fore he left in his blue box,
Pressed but a little gift into my hand.
Quoth he: "If George is fagged after the fight –
And who could blame the old boy if he was? –
Then place this extra battery in his pack,
He'll find the strength, at least, to take applause."
Then added he: "Zadek, I have to go;
I pulled the Randomiser out of joint
To bring the TARDIS back to Iara's shores,
And now I sense disturbance in the air –
That old Black Guardian is on my tail."
And thus he went, with great celerity,
With flashing light, and wheezing, groaning sound.
The birds flew from the trees as he rose up.

FARRAH

Wise Doctor. He hath saved our country twice.
Once from the Count of hateful memory
And now from foes arrived from the stars.

ZADEK

And when Coriolanus' strength has gone,
There'll be another battle lost and won.

ALL

To Coriolanus come all joy and honour!

Flourish of cornets. Exeunt all.

Is this a (Dalek) I see before me?

MASTER FAUSTUS

One of the more extraordinary inclusions in the Shakespeare Notebooks is this extract from a play entitled Master Faustus. *On the face of it, this seems to be either an early draft or a reworking of the play* Doctor Faustus – *which was written, of course, by Shakespeare's contemporary Christopher Marlowe. Or was it? Did Shakespeare include in his Notebooks material actually written by Mawlowe, or is this proof that Marlowe himself derived his own work from a previous text by Shakespeare?*

Whatever the truth, this extract makes for fascinating reading – not least for the inclusion of Marlowe as a character within the drama. His death has some resonance with the actual event, though with its inclusion of 'Daleks' (presumably evil spirits) this is evidently intended as a 'fantasie'.

SCENE I – A TAVERN IN DEPTFORD

Enter MARLOWE, a playwright. He is accosted by DOBBIN and DULLBERRY, two ruffians.

DOBBIN
  Prithee, art thou the famous playwright?

MARLOWE
  Ay.

DOBBIN
  Fellows, 'tis Master William Shakespeare!

MARLOWE
  Shakespeare?

DOBBIN

The greatest writer who ever did draw breath!

DULLBERRY

Romeo and Juliet!

DOBBIN

Ye Comedy of Errors!

MARLOWE

I wrote not them. I am Christopher Marlowe.

DULLBERRY

Morley?

MARLOWE

Marlowe.

DOBBIN

Who?

MARLOWE

Know thee not the Tragedie of Tamburlane?

DULLBERRY

I fear me not.

DOBBIN

Though, I am full sure it is most good.

DULLBERRY

If not as good as those of Master Will. Hast thou met him?

MARLOWE

I know him well.

DOBBIN

Then tell him how good we think him.

MARLOWE

I shall.

DOBBIN

He really is very good.

DULLBERRY

Truly. No hack poetaster he.

MARLOWE

Friends, for so I call ye, I know right well your intention.
You art the Lord Keeper's men, are ye not?
Come to pay me my bloody recknynge.
But the sun is shining, and I shall not fight with you today.

DOBBIN

He turns down our challenge?

DULLBERRY

He doth. The insult shall not stand.

They draw and fight. Enter MAGISTER.

MAGISTER

Put up thy bright swords noble gentlemen,
Else the dew will rust them.
For I am the Master and thou wilt obey me.

DOBBIN AND DULLBERRY freeze.

MARLOWE

I am impressed.

MAGISTER

Christopher Marlowe, come with me.
I have an offer you can not refuse.

SCENE II – MARLOWE'S HOUSE

MARLOWE

What manner of man are you, sir?

MAGISTER

I am a man of stars, who has waited in the skies
Long have I wished to help you, but I feared
I should overwrack your mind.

MARLOWE

A spirit?

MAGISTER

Let's call it that. And so to work.

MARLOWE

What is your business with me?

MAGISTER

Good Marlowe, but glance close through my tricked glass
I have a thing to show. A vision of a scribe
Mighty, all garlanded with poesy's bays
And right clapper-claw'd to infinity
Time has no measure, nor does thy fame.
One name shall echo cross the stars. Marlowe.

MARLOWE

Say you so?

MAGISTER

Ay. I am the Master. My word's my power.
And Kit, I give it thee. Say what you see?

MARLOWE peers through MAGISTER's magical scrying glass

MARLOWE

    I am amazed. I but hoped my words would
    Outlive my span a little. But here I see them
    Printed, studied, acted, quoted, learned
    Quite picked apart and then repatched
    Brought to life and done to death unending.
    I see my plays acted through the ages.
    I see boys squeak my lines in endless generation,
    In dumb show processional til the crack of doom.
    I see a stage lit by dying stars, and
    On it voic'd my words, their final echo
    Cloak'd by the closing curtain of creation.
    I see only Marlowe, Marlowe, Marlow. My unending
    Line. You, Magister, you have shown me
    The book of time, and my whole life within.
    I should be a dot, a blot, the flea's flea.
    And yet, writ large, just MARLOWE still I see.
    Do you lie?

MAGISTER

    Lie? I never lie.
    I fear my vision has much shook thy brains.
    Good Marlowe, lay that damned book aside,
    And gaze not on it lest it tempt thy soul.

MARLOWE

    Nay, I'll see more. Christopher Marlowe?
    A Canterbury cobbler so oversouled?
    With nature's pride and richest furniture
    My works do menace heaven and dare the gods!
    And yet, there is a name I thought to see
    Writ on eternity's fair brow. I squint
    But see it not. My Will is Shaken by it.

MAGISTER

> Oh, Master Shakespeare? Fear him not.
> His bright eyes suddenly burn so pale.
> For he is but a candle in the wind.
> And I shall snuff him out.

MARLOWE

> I will no harm to Will. And yet –

MAGISTER

> And yet.

MARLOWE

> Yet. Spirit, good or bad, why help you me?

MAGISTER

> I am one who weaves behind the rich
> Tapestry of time. I pick up threads,
> Pull colours, endeavour and much mend.
> You are one gold yarn I seek to lengthen.
> I do prepare a trap beyond a trap
> Which one day it would be a joy to spring.
> Against that day I need you, Kit.
> I can't get you out of my head
> You work is all I think about.
> For there is a dark secret in you.
> Forsake thy king and do but join with me
> And we will triumph over all the world.
> I hold the fates fast bound in iron chains
> And with my hand turn Fortune's wheel about
> May we become immortal like the gods.

MARLOWE

> Had I as many souls as there be stars
> I'd give them all my Mephistophilis
> By you, I'll be great conqueror of the world
> And make a bridge through the moving air.

SCENE III – MAGISTER'S LIBRARY

MARLOWE enters, much amazed

MAGISTER

    I do nothing Marlowe, but to delight thy mind
    And let thee see what magic can perform
    Hold, take these books, peruse 'em thoroughly
    The iterating of these lines brings gold.

*Pageant: As Marlowe reads, figures dance about him in
merry processional*

MAGISTER

    What think you, Kit? Have I not shown you
    All the greatest stories of the world?

MARLOWE

    I have solved the code of great Da Vinci
    I have beheld all fifty shades of grey
    And met Dame Bridget who counts all her food
    And learned the secrets of that galaxy
    So long ago and far, far away.

MAGISTER

    And were you struck?

MARLOWE

    Well,
    I liked the strong magicks of young orphan
    Harry. That is a story which methinks has legs.
    And yet…

MAGISTER

    And yet?

MARLOWE

    Yet.

MAGISTER

> Fear not. I have yet wilder skies than these.
> I call down heavens, unwrap the stars
> He must needs go that the devil drives
> There shall be no rest for the wicked
> And no sleep till Brooklyn.
> I will be Paris, and for love of thee
> Instead of Troy, shall Skaro's towers be sacked
> Come, step into my magic cabinet.

> *At MAGISTER'S invitation, MARLOWE and he*
> *enter a box*

SCENE IV – MAGISTER'S VAST CABINET

MARLOWE

> What great reckoning in a little room!
> Is this the jakes of Ajax? This box is
> Bigger on the inside than the outside!

MAGISTER

> You have seen nothing yet, my friend.

> *MAGISTER performs a conjuration.*
> *Infernal trumpets sound.*

MARLOWE

> What dread noise is that?

MAGISTER

> 'Tis the music of the spheres.
> They are the alarums of our excursion.
> See? We ride the back of time.

From Peru to Cebu, hear the power of Babylon,
From Bali to Cali – far beneath the Coral Sea.
We sail away, sail away, sail away.

MARLOWE

I am amazed, gentle Mephistophilis.
Our souls whose faculties can comprehend
The wondrous architecture of the world,
And measure every wandering planet's course
Still climbing after knowledge infinite
And always moving as the restless spheres
Will us to wear ourselves and never rest.

MAGISTER

We are landing. 'Tis but the gentlest bump. Come let us go.
But Kit, beware. We voyage on a planet of total war.

SCENE V – THE DEAD REALM OF SKARO

Enter a pageant of DALEKS

DALEKS

We sing in praise of total war
Against the Thals whom we abhor
To free the tomb of Zeg our lord
We'll put all creation to the sword
There is no greater glory than
To burn with fire the lake of Darren

The DALEKS espy MAGISTER and MARLOWE

DALEKS

Do not move! Do not move! Exterminate!

MARLOWE

What wild mechanicals? What dread armour!
I will not fear bugbears and hobgoblins
And utterly scorn both gods and monsters.

MAGISTER

Your boldness does you merit. Yet should we run.
Here have I left a little work undone.

DALEKS

Halt! Thou are the Master! Exterminated shalt thou be!

MAGISTER

Run.

MARLOWE

Run?

MAGISTER

Run!

They flee back to Magister's Cabinet

SCENE VI – MAGISTER'S CABINET

Enter MARLOWE and MAGISTER at a fast pace

MARLOWE

If heaven were made for man, 'twas made for me
Have I not made music with my Mephistophilis?

MAGISTER

I think there are many stories here for you.
All the roads that lead you there are winding
And all the lights that see you there are blinding
But after all, thou shalt bow to my wonder will.

MARLOWE

How know you of these worlds and creatures?

MAGISTER

As you have Will, I have my adverse,
A dreadful Doctor, a trickster japery
A gallivanting gallimaufry of Gallifrey.
He stirs these creatures up like wanton boys
Sticking a hornet's nest, and 'tis my job
To clear up his mess. We are Lords Temporal
Falling in endless fight like Lucifer and Gabriel
Our tales entwined. My destruction eternal.

MARLOWE

O what a cozening Doctor was this to practice on you so!
Where next, gentle Magister?

MAGISTER

We'll chase the stars from heaven and dim their eyes
That stand and muse at our admired arms.
We'll crest fair the moons of frozen Telos
And loop the lonely tail of Mondas, quick
We are but twenty-four hours from pulsar.

Pageant of stars unfurled

MAGISTER

We cross the void beyond the mind,
The empty space that circles time
We see where others stumble blind
To seek a truth they never find
Eternal wisdom is my guide
I am the Master.

MARLOWE

But is there no tale compassed in firmament of Earth?
Lower my horizons from Jove's thunders.

MAGISTER

Fair enough, Kit. Thou art a small town boy
And this is a crazy world. Let us now
Talk not of paradise or creation, but mark the show
What Rome contains for to delight thine eyes?

Pageant – enter CAESAR. He is stabbed by
CONSPIRATORS. Following him in procession
are many famous ROMANS

MAGISTER

The noblest Roman of them all, cut down.
And here is Cleopatra coming at you,
Age cannot wither her, nor custom stale
Her infinite variety. She's beautiful
She's beautiful indeed.

Enter more characters

MARLOWE

And what means the sour looks of this sad prince?

MAGISTER

This is tragic Hamlet, Prince of Denmark
'Tis a mad world, 'tis a sad world for him.
Followed by Scotland's dark King Macbeth
Here come massy Faerie throngs
And there prosperous Prospero.
Shall not I make dour Hamlet sing to thee?
Here's Helen, the face that launched a thousand ships
And burnt the topless towers of Ilium.

MARLOWE

All! All! These people cram up my brain with good ideas!
Settle thy studies, Marlowe, and begin
To sound the depth of that thou wilt profess
Be a playwright, Marlowe, heap up gold,
And be eternized for these wondrous parts.
Couldst thou make men to live eternally,
Or, being dead, raise them to life again?
No emperor shall live but by my leave.
What doctrine call you this?

MAGISTER

Che sera sera
What will be, will be. Divinity adieu
Here Marlowe, try thy brains to gain a deity.

MARLOWE writes quickly as the pageant dances

MARLOWE

I have a thought.

MAGISTER

You have many.

MARLOWE

Yet more.

Tell me, noble Mephistophilis, where

Sprang these great stories that fill my mind?

MAGISTER

Ah, ask not that.

MARLOWE

I must and shall.

MAGISTER

When thou took'st the books

To view the stories, then I turned the leaves

And led thine eye. Shall we say

Where there's a will, there's a way?

MARLOWE

Shakespeare?

MAGISTER

All his fine plays are laid before thee.

His labouring brain

Begets a world of idle fantasies.

At risk of spoilers, I confide to thee

These are all the works he has yet to write,

And every one a winner.

MARLOWE

Quod me nutrit, me destruit

What feeds me also brings me down.

Magister, what have you done? I am undone!

MAGISTER

Say it ain't so. Surely better the devil you know.

This, here I swear by my royal seat.

MARLOWE

Well, you may kiss it then.
O would I had never seen London, never read book
Oh for the vain pleasure of four and thirty plays hath
Marlowe lost eternal joy and felicity.
False friend thou. My pen is overcast,
It inks another's brains. Forgive me, Will.
Why this is hell, nor am I out of it.

MAGISTER

Ungrateful wretch! I'll make thee doll small.

MARLOWE

Fine. Lay on, sir. Magister do thy worst.
All these glorious stories shall still be told.
But not by me. I'll write no more.
Ah, I'll burn my books.

MAGISTER

You'll write no more?

MARLOWE

Not one word.

MAGISTER

Hadst thou kept on that way, Marlowe behold
In what resplendent glory thou hadst sat
In yonder throne, like that bright shining Shakes.
Tea towels, mugs, placemats and penguin classics
But, no, you suffered a loss of faith, and so must
Your stories end.

MARLOWE

Come not near me, Magister.

MAGISTER

Come, gentle coz. Pick up they pen.
Thou art bewitched, bothered and bewildered.
Just one play more – Othello, why he's a jolly fellow
Or Sir John Falstaff, he's good for a laugh.

MARLOWE

I'll pen no more. My stop is full.
If all the pens that ever poets held
Had fed the feelings of this Master's thoughts
If all the heavenly quintessence that they still
From their immortal flowers of poesy
Yet should there hover in their restless breasts
One thought, one grace, one wonder, at the least
Which Into words no virtue can digest.

MAGISTER

Sometimes I find thee hard to follow. Does this mean you're done?

MARLOWE

Ay.

MAGISTER

Then hear that drum?
'Tis the sound of the underground
The beat of the drum goes round and round.
These boots were made for walking
And that is just what they shall do
Today my boots shall walk all over you.
Go back to thy tavern. I'll not stop you.
Your appointment with fate is overdue.

MARLOWE

Aye, I have done her majestie good service
Under cloak and dagger, dagger and cloak.

I have few friends at home, many enemies abroad
I have long deserved to be rewarded for my faithfull dealinge.
Return me to the inn where you found me
My recknynge must at last be paid in full.

MAGISTER

I could have saved you. It is not yet too late.
'Tis your last chance of living and your first chance to die.
I can yet keep you from the last chance saloon.

MARLOWE

Nay good close friend, for so I call thee Magister.
O gentle Marlowe, leave this damned art
This magic, that will charm thy soul to hell.
The stars move still, time runs, the clock will strike
The devil will come and Marlowe must be damned.

Exeunt

MAGISTER

Say thou won't leave me no more
And I shall take thee back again
I shall forgive and forget
If thou say thou shalt never go
For 'tis always better the devil you know.

SCENE VII – A TAVERN IN DEPTFORD

MARLOWE

Hulloa Dullberry, Adieu Dobbin
Spials all. Unsheath thy curtle-axes
For there sits Death, there sits imperious Death
Keeping his circuit by the slicing edge
Fall to friends.

Master Faustus

They slay MARLOWE

MARLOWE
My bloodless body waxes chill and cold
And with my blood my life slides through my wound
My soul begins to take her flight to hell
And summons all my senses to depart.
This is my mind and I will have it so.

Dies

MAGISTER
Ah well. You can't win them all.
Pride comes before a fall.
I have played all my cards
And that is what you've done too
Besides victory, there remains destiny.
The loser stands right small
And the winner takes it all.

*Double, double, toil and trouble;
if we are caught in a time bubble.*

THE SONNETS

These early drafts of some of Shakespeare's well-known Sonnets appear throughout the Notebooks. Where they are prefaced by a dedication or other explanatory text, this is given. But most are simply presented as they appear here.

Sonnet 1
This version of the sonnet appears in the Notebooks addressed, a MS addition asserts, 'To an Old Man by his Grand-daughter'.

From fairest creatures we desire increase,
That thereby beauty's rose might never die,
Those Venus night-fish cooked on a caprice
The plants of Esto when they sing and sigh.
I have of late seen Daleks in the street,
In Bedfordshire and on the Edgware Road.
And though they suffered terrible defeat,
Here thrive the weeds their seeds of war have sowed.
To mine eye, though, this seems not the world's end.
'Tis but a new one moving within reach.
Rivers unchoke, skies clear and deep wounds mend,
For here dwells my belov'd. Thus I beseech:
       Go forward in beliefs I know are thine
       To prove that I am not mistook in mine.

Sonnet 2
This sonnet is apparently 'From a Lord departing into Exile'

When fifty winters shall besiege my brow –
Though add a thousand more to get it right –
Yes, then, and only then, could I allow

You lot to send me into that good night.
I stand here in the dock before a screen
On which you've thrown a hideous display.
The most peculiar bunch I've ever seen:
Too old, too young, too fat, too thin. I say,
Is keeping my own head too much to ask?
If I'm confined in exile in one place
Defending it from monsters – onerous task! –
It's better done with a familiar face.
      I do maintain I have the right to choose!
      Oh dear. That nose! (And what's with these tattoos?)

Sonnet 3

Again, this early version of what would become Sonnet 3 is prefaced in the Notebooks by a note: 'From a friar to his pupil'.

Look in thy glass and tell the face thou viewest
Now is the time that face should form another;
Whose fresh repair if now thou not renewest,
Will fall past Artron's power to recover.
For thou hast trod this troubled way before:
Worn thin by Mondas at the snowcapped pole;
Redrawn by force in Time Lord court of law
And cast as army helpmeet, or such role.
But now, thou needs must face thy greatest fear.
The Great One's fire hath hollowed out each bone,
And death, or death's own spider creeps too near
For thou to flee. 'Tis time to die alone.
      Old man, embrace the joy of thy removement.
      Methinks the nose a definite improvement.

This was the noblest Romana of them all.

Sonnet 12

When I do count the clock that tells the time
And see the brave day sunk in hideous night,
When I behold the readings' ceaseless climb.
And mark the Fault Locator's flick'ring light
When scissor blades are sunk into the bed
Tight hands around my neck increase their grip,
And on the scanner images are read
Of Quinnis where we nearly lost the Ship
Then of thy motives do I question make,
And think to banish thee into the void
If not for mine, then for my Susan's sake,
As soon, it seems, we four will be destroyed;
        Unless some other cause I can construe
        Within these shapes of twisted ormolu.

Sonnet 14

From lights above I did my judgement pluck;
Now I possess a new Astrology,
But not to tell of good or evil luck,
Of plagues, of dearths, or seasons' quality;
My prophecies have changed their nature since
A light fell from the heavens into mine eyes
Now I do steer the fortune of a prince,
Determine which duke lives and which king dies.
There lives now here on Earth a fateful star
It burns among my acolytes, and soon
The fateful helix called Mandragora
Will swallow up the pale face of the moon
        For San Marino I prognosticate:
        A holy darkness falls upon this date.

Sonnet 15

When I consider everything that grows
Beneath the span of my cathedral green
The liberated bonsai trees, the rose,
The crabbed old pod no botanist has seen.
When I perceive that men owe their increase
To felling forests, gorging herbs and fruits;
Despoiling precious verdure without cease
Tearing defenceless nature at the roots,
I am compelled to sow their seeds of doom,
To found a green and silent paradise:
O, let a million hungry flowers bloom
Inspired by life from 'neath Antarctic ice.
      A revolution blossoms in these grounds
        My *Floriana Requiem* resounds.

Sonnet 16

A note prefaces this sonnet: 'From an Exile to his Dead King'.

But wherefore do not you a mightier way
Make war upon this bloody tyrant, Time?
And fortify your self in your decay,
As I did, 'tombed within Jurassic lime?
For years thrice fifty million I endured
Sustained by hope of vengeance on my kind;
A fragment lost beneath the earth, abjured
By all the judges that I left behind.
And here you sit, a corpse upon a chair –
Old Rokon, monarch of an empty waste
Mummified in Kastria's glacial air;
Cities and thrones and powers all quite erased.
      So I, as next in line, you now address
        As I become the king of nothingness.

Sonnet 18

Perhaps the best-known of all Shakespeare's sonnets, this early draft appears without explanation.

Shall I compare thee to a Type Fifty?
Thou art more lovely and more temporal:
Rough time winds shake the positronic flow,
And Fast Return hath all too short a spring:
Sometime too hot the Eye of Harmony
Is by a Temporal Orbit stopped at last
And every wheezing groan sometime declines,
By chance, or Vortex changing course untrimmed:
But thy materialisation shall not fade,
Nor lose possession of thy Time Rotor,
Nor shall death brag thou wander'st Gallifrey,
Wherein eternal Rassilon dost thrive,
      So long as Time Lords plot, or Daleks kill,
      So long my TARDIS will you serve me still.

Sonnet 19

Devouring Time, blunt thou Urbankan claws
And make Xeriphas' brood devour its own.
Pluck the keen teeth from dreaded Mara's jaws,
And bring the baleful Malus crashing down;
Shipwreck the chill Eternals and despoil
Whole fields of celery and orchids black.
Let me pick chacaws or endure the toil
Of heaving oars for monstrous Captain Wrack.
But I forbid thee one most heinous crime:
Let not Spectrox extract its deadly due
From Peri, newly travelled into time.
Until I prove Professor Jackij true
      I'll struggle on until my final breath.
      Feels different this time. Brave heart. Is this Death?

Sonnet 27

Weary with toil, I haste me to my bed,
The dear repose for limbs with travel tired;
But then begins a journey in my head
To work my mind, when body's work's expired:
Methinks I stood upon a high white tower
Fell like a stone towards the grassy earth;
Methinks I felt an occupying power
Polluting me; disrupting my rebirth.
Two faithful friends brought me from outer space
To Castrovalva, shining on a hill.
It seems to be tranquil, ordered place,
And yet these painful dreams, they plague me still.
      A shadow now falls fast, it seems to me,
      Upon the dwellings of simplicity.

Sonnet 29

Again, there is an introductory note to this early draft: 'The lament of a Great Engineer'.

When in disgrace with fortune and men's eyes
I all alone beweep my outcast state,
And trouble deaf heaven with my bootless cries,
And look upon myself, and curse my fate,
Wishing that I had found some safer course
To gift my race with temporal power undreamed.
When I absorbed that nova's dreadful force
I burnt within the sun. I died, it seemed.
But now I know that firestorms worse than hell
Corroded me to something like a ghost.
And in this will-created world I dwell
Denied the life that I desire the most.
      They think of me, and my achievements laud.
      But this I trow: I should have been a god.

Sonnet 53

What is your substance, whereof are you made,
That millions of strange shadows on you tend,
Black multitudes concealed in every shade?
A death too horrible to comprehend
Is met by those who stray into your maw.
You are the twilight settling on the stair;
The darkling form upon the marble floor;
Who would have guessed that motes upon the air
Could harbour such an appetite for flesh?
Methinks I see you now, between the shelves;
A cloud of hunger, seeking something fresh.
A dread thought chills my heart. Between ourselves,
> There is no fate discernible to me.
> Except a silence in the library.

Sonnet 116

Let me not to the marriage of true minds
Admit impediments. Love can rise fast
When Vardans and rough beasts of other kinds
Suggest that ev'ry moment is your last.
But true love is a fix'd co-ordinate
Measured from Galactic zero centre.
From one firm course it does not deviate
Well, that's the view of your erstwhile mentor.
Love's not Time's fool, though husbands of this sort
Will not keep the same face from year to year.
The life of Sevateem might seem too short
Compared with those of Gallifrey, I fear.
> If this be error, then it must be mine.
> So pay no heed, and instead, ask K-9.

Sonnet 123

No, Time, thou shalt not boast that I do change:
The pyramid that forms my prison here
To me is nothing novel, nothing strange.
Millennia immobile on this bier
Have clarified my mind and cooled my blood
With but a single thought: Where'er I tread
I leave dust and darkness. I find that good.
My brother Horus left me here for dead;
But far from dead, as death would be too sweet.
The vengeance of Osiris was too cold
To offer better terms for my defeat.
And thus I sit and wait as stars grow old.
       But this I vow and this shall ever be;
       All life – fish, fowl, reptile – is foe to me.

Danger knows full well,
the Doctor is more dangerous than he.

AS YOU LIKE IT

This transcript from an early seventeenth-century staging of the play differs considerably from all known sources of the text. There are obvious similarities, so the diversion may have been unique to this performance, perhaps an extreme example of improvisation by the actors involved.

Duke Senior has been usurped by his brother, Frederick, and banished from the court to the Forest of Arden. Among his fellow exiles is the melancholy Jaques. And they are not the only party obliged to make a new home in the wilderness.

JAQUES
    A fool, a fool! I met a fool i' the forest,
    A motley fool; a miserable world!
    As I do live by food, I met a fool
    Who roved from grove to grove, as if in search
    Of quarry lost upon those verdant grounds;
    Some prize he thought to snatch from out the air.
    "Good morrow, fool," quoth I. "O, sir," quoth he,
    "Has't seen a press abandoned in these woods?"
    And then he drew a dial from his poke,
    And, looking on it with severest eye,
    Says very wisely, "I set up the HADS
    To move the girl upon this very plot.
    Methinks the course co-ordinates hath slipped
    And sent her to a place that I know not.
    And so from hour to hour she drifts and drifts
    And I from place to place must make pursuit.
    And thereby hangs a tale." When I did hear
    The motley fool thus moral on the time,
    My lungs began to crow like chanticleer,

That fools should be so deep-contemplative,
And I did laugh sans intermission
An hour by his dial. O noble fool!
A worthy fool! Motley's the only wear.

DUKE SENIOR
What fool is this?

JAQUES
O worthy fool! One that hath been a lord,
And speaks of grave disruptions in the sky.
Strange tales of wounds within the walls of time.
And signs unread in any almanac.
I wonder'd at his humour: in his brain,
Which is as dry as the remainder biscuit
After a voyage, he hath strange places cramm'd
With observation, the which he vents
In mangled forms. O that I were a fool!
I am ambitious for a motley coat.

DUKE SENIOR
Thou shalt have one.

JAQUES
It is my only suit. A coat composed
Of hues that would perforce to boil
The jelly in thy orisons.
A dial within my poke that sings
Like crickets sealed within a stoup of wine.
And liberty to roam where'er I please,
Discourse cosmology unto the trees.
Invest me in my motley; give me leave
To speak my mind, and I will through and through
Cleanse the foul body of the infected world,
If they will patiently receive my medicine.

IN ANOTHER PART OF THE FOREST

Enter PERPUGILLIAM, with a paper, reading

PERPUGILLIAM

From Dunsinane to Windsor town,
No jewel is like Peri Brown.
Her worth, like pearls set in a crown,
Processes forth as Peri Brown.
Let no man curse or wear a frown
Within the orbit of Miss Brown.
The shipwrecked sailor, though he drown,
Dies glad with thoughts of Peri Brown.

CLOWN

Glad? Glad? Glad? I never heard so egregious a verse. A paper
bearing execrable work has only one fit function. I shall not
elaborate.

PERPUGILLIAM

Out, fool!

CLOWN

For a taste:
She looks alluring in a gown
But leotards love Peri Brown.
Each adjective must have its noun:
So loud and shrewish go with Brown.
She runs, she screams, and then falls down
And so do all with Peri Brown.
She flies through space, she's been around
O by my stars! 'Tis Peri Brown.
This is the very false gallop of verses: why do you
infect yourself with them?

PERPUGILLIAM

Peace, you dull fool! I found them on a tree.

CLOWN

Truly, the tree yields bad fruit.

PERPUGILLIAM

Who is't that put these apples on the bough? These purple
lines on little squares of blue, that deck the branches here
in Arden's wood? Can'st not be a lover, for 'pon our travels
in the void the ones who lavish me with their amours are
seldom men one could take home to court. Rather, they
creak'st in piebald leather, or sit enthroned in darkness, wet-
lipped, observing me upon the glass.

CLOWN

You have said; but whether wisely or no, let the forest judge.

PERPUGILLIAM

O, here's another. Stand aside.

CLOWN

Too late, I have it.
[Reads] Why dwell we in this desert here?
Have we betrayed our mistress? No:
From our true course we did not veer
And yet we are marooned so
Upon some bank of time beyond the trees.
We sees't thou, as sparrow spies the crop;
But, borne like thistledown upon the breeze,
We mount the wind, and sail its giddy top.
O Perpugilliam, in Arden's maw
Draw down thy helpless servant from the skies.
A sight funereal as e'er thee saw
Is high above you, hidden from thine eyes.

A friend enshrouded in the veils of space
That longs for harbour in another place.
But here we must remain in bondage aerie,
Until we once again find sight of Peri.

PERPUGILLIAM

I was never so be-rhymed since Pythagoras' time, that I was a
Varosian parrot, which I can hardly remember.

CLOWN

Trow you who hath done this?

PERPUGILLIAM

Is it a man?

CLOWN

More than a man.

PERPUGILLIAM

I prithee, who?

CLOWN

Is it possible?

PERPUGILLIAM

Nay, I prithee now with most petitionary vehemence, tell me
who it is.

CLOWN

Our home, our berth, our star-crossed ship of time.

PERPUGILLIAM

O wonderful, wonderful, and most wonderful wonderful!
And yet again wonderful, and after that, out of all hooping!

CLOWN

Received with thanks. Now take this in thy hand.

PERPUGILLIAM
What is't?

CLOWN
It filters feedback on the sub-etheric band
And sends a signal soaring through the spheres.
O, list. It is the TARDIS. She appears.

A great wheezing and groaning, below. Exeunt.

'Tis now the very witching time of night,
when things get really exciting and dangerous.

DOUBLE FALSEHOOD

Drawn together from a variety of sources, the materials reproduced here pertain to Double Falsehood – *which purported to be a version of Shakespeare's play* Cardenio. *Unfortunately the only source for the final newspaper article is incomplete in the British Library.*

Readers are invited to draw their own conclusions as to the veracity of the text.

ALEXANDER POPE'S DIARY

14th Dec, 1727

Saw "Double Falsehood" (aptly called) lst nght, put on by the wretched Tibbald. Claims to be an adaptation of Shakespear's "Cardenio", but an obvious forgery. The graveyard turf of Stratford must be churned up with shame.

THE COFFEE CHAT NEWSPAPER

January, 1728

Happening to come across Mr Theobald, late of the success of "Double Falsehood", I was able to corner him about his play, which he claims to be from the pen of Shakespeare.

"Oh it is, it is," he vowed, colouring. "The only claims against it rise from Mr Pope, who slights against it regularly. His raillery is boundless."

Boundless? Are not some of the lines perhaps a little less than the Bard would grant?

"Well, that they are, if so they were. But Mr Pope wilfully misquotes them to make them nonsense."

Why would he do such a thing?

"Revenge." Mr Theobald colours still further. "He took against my few friendly critical notes on his edition of Hamlet."

Ah yes. Mr Theobald refers to the 200 closely printed pages of criticism he published, which caused Mr Pope to fly into a fury. Is it not true that Mr Theobald plans his own edition of Shakespeare?

"Indeed." the author puffs himself up, "And it will most certainly contain Cardenio."

But how did the missing play come into his hands?

"I have three manuscripts for the play, actually. My adaptation uses material from all of them, but the original work is perhaps better read than performed. Hence my mild modern adaptation to the current sensibility."

So, it seems Mr Theobald is keeping not one, but three copies of "Cardenio" close to his chest. Until then, we must attend Drury Lane and try and glimpse the Bard through the noble Theobald's shards.

THE HISTORY OF CARDENNA

Sir, I went along last night in order to take down a shorthand report of the play for printing, but alas, afterwards found my notes had been corrupted by the speech of the playgoers sat next to me. I can attend the next performance, if you'll pay extra?

HENRIQUEZ
She weeps, be gentle to her good Barnardo.

LEONORA
Then woe the day! I'm circled round with fire.
No way to escape but through the flames.

THEOBALD
Oh, raptures!

POPE
Pah!

DOCTOR
Shaddup Alex and eat your popcorn.

POPE
Sit through more of this, Doctor? I'd sooner eat spoons.

THEOBALD
But it's just how I'd imagined it, Mr Pope.

POPE
You, Theobald? An imagination!

DOCTOR
Someone is not getting his share of Vienetta on the journey home.

DON BERNARD
Camillo, pocket up your indignation
And attend the truth of your fair daughter.

POPE
Can I get some more wine? I've finished this glass. Actually, just pass me the bottle.

THEOBALD
Actually, could I have a drop –

POPE
No!

DOCTOR
Some people. I dunno. I bring the two of you back in time to see Shakespeare's "Cardenio". Some people would be grateful.

THEOBALD
I am, I truly am!

POPE
Course he is. The idiot would be happy licking a salt stick.

THEOBALD
Hey!

POPE
What a wasted opportunity. We could have attended the first night of "Hamlet" to resolve what was the proper text.

THEOBALD
Like you'd know what that was if it bit you, you tiny fraud.

DOCTOR
And it's all the Viennetta for me.

CITIZEN
Though that a thousand daggers barred my way
I'd dare em all to serve you.

POPE
Mr Theobald, I owe you an apology when I accused you of forging this play.

THEOBALD
I humbly accept it, Mr Pope.

POPE
Accusing anyone of writing this nonsense is a vile insult.

They fight.

DOCTOR
Fellows, fellows, stop scrapping and hush. Can't you see the chap next to us? He's pirating the performance.

THEOBALD
For publication?

DOCTOR
Yes. Shush.

POPE

But there's no such pamphlet.

DOCTOR

Of course not. You're talking all through it. You're ruining the transcript. Hello, sorry about my friends. They're really very nice. Terrible taste in clothes and are you taking down every word I say? Oh yes, I see you are. I'll shut up now. This is really very distracting for you, isn't it? No, honestly, shutting up. Zip. Shtum.

JULIO

Poor Leonora! Wretched, damned Henriquez,
She bids me fill my memory with danger.

POPE

Eleven syllables? I've seen more ordered poesy in a sailor's brawl.

THEOBALD

Like there are docks in Twickenham, toad. Ow!

POPE

I've another finger to poke your other eye. There's plenty more where that came from.

THEOBALD

You're doing this deliberately, aren't you? Now you've realised that the fellow next to us is taking down the playscript, you're doing what you can to thwart his transcript so that "Cardenio" doesn't survive.

POPE

What if I am? If a printed text already survived, how could you then claim to rediscover the play? What would you base your meagre fortunes on then?

THEOBALD

I would just be happy for the world. Shakespeare belongs to the generations.

POPE

Hah! You're just after a fat editor's fee for your Complete Works.

DOCTOR

Ladies, hush now, or you're walking back to the 1730s.

POPE

But we're getting to the good bit.

DOCTOR

Oh, so you're enjoying it now, are you?

POPE

Yes. It's the worst line of Tibbald's entire forgery.

THEOBALD

It is not a forgery. I've shown through minute analysis how such a line –

POPE

"None but itself can be its parallel
And by such a friend profess'd"? As if Shakespeare would bother with such bumfodder.

JULIO

Is there a treachery like this in baseness?
None but itself can be its parallel
And by such a friend profess'd!

POPE

Oh.

THEOBALD

See? I was right. You owe me an apology.

POPE
  Doctor?

DOCTOR
  Hmm?

POPE
  Why is your transcriber friend looking at us so strangely?

THEOBALD
  Is it because it's the first night?

DOCTOR
  Oh, oh dear. Helloooo – would you mind ignoring us and just concentrating on the historical first night of "Cardenio"? There's a slice of Vienetta in it for you.

THEOBALD
  I think you owe me an apology.

POPE
  Whatever for, my dear sir?

THEOBALD
  You told all London this play was forged by me. You had the performances catcalled, my edition of Shakespeare cancelled. Because of you I live in poverty.

POPE
  My good fellow, I did nothing of the sort. I may have dropped the tiniest of hints at the coffee house. That is all. You really are being over-sensitive.

THEOBALD
  You owe it not just to me, but to literature to recant. If you support me, then I can publish the three manuscripts of "Cardenio" I have –

POPE

Three, you say?

THEOBALD

Aye. Stored at the Covent Garden Theatre. So, anyway, if you were to –

POPE

Yes, yes, of course. I'll absolutely apologise. More than that, my dear sir, I'll dedicate my next poem to you.

THEOBALD

I misjudged you.

POPE

And I you. The least I can do is to afford this play the attention it deserves. And now, if you'll excuse me, I just need to visit the close-stool.

JULIO

Hold out thy faith against the dread assault
Of this false lord.

DOCTOR

Where's he gone?

THEOBALD

The jakes.

DOCTOR

Really? That's funny, he bumped into me as he – wait, the key! The key to my ship is gone. The little tinker, he's clearing up the Vienetta.

THEOBALD

It's worse than that, I fear.

NEWSPAPER DATED 1808

COVENT GARDEN THEATRE DESTROYED BY FIRE

A mysterious fire swept through the heart of London's theatre district last night. Covent Garden Theatre was destroyed before the blaze could be controlled, along with the Theatre Museum in the basement. It is reported that a strange, imp-like figure was seen darting into the building shortly before the conflagration.

"We've lost playbills, so much of the rich pageant of England's historic stage," said Mr Humphrys, the manager. "It's impossible to work out the loss. For instance, we lost all three copies of the play that Mr Theobald based his Double Falsehood on. Which many people claim was written by

Here the extract ends abruptly, the rest of the page being missing. It is interesting to speculate on how it might have continued, but the loss of material is, to say the least, frustrating.

There is a special providence in the fall of a sparrow.
And a Dalek mothership can spot the fall
of a sparrow from geostationary orbit.

HAMLET

This transcript of Hamlet *(Act V Scene i) survives from the first performance, and shows some interesting, not to say enigmatic, variation from the accepted text.*

GRAVEDIGGER
Here's a skull now. This skull hath lien you i' th' earth three-and-twenty years.

HAMLET
Whose was it?

GRAVEDIGGER
A whoreson mad fellow's it was.

Enter a CLOWN, guised most strangely in great coat, long scarf of rainbow colours and brimmed hat.

CLOWN
Well, whose do you think it was?

HAMLET
Nay, I know not.

CLOWN
I'm not surprised. He's not looking his best just now, is he?

GRAVEDIGGER
A pestilence on thee for a mad rogue! 'A pour'd a flagon of Rhenish on my head once. This same skull, sir, was Yorick's skull, the King's jester.

HAMLET
This?

GRAVEDIGGER
E'en that.

CLOWN
You don't say.

HAMLET
Let me see.

Takes the skull.

Alas, poor Yorick! I knew him, Horatio.

CLOWN
Oh everyone knew poor Yorick. Dreadful teeth, I seem to remember.

HAMLET
A fellow of infinite jest, of most excellent fancy. He hath borne me on his back a thousand times. And now how abhorred in my imagination it is! My gorge rises at it. Here hung those lips that I have kiss'd I know not how oft.

CLOWN

Well, you can't be expected to remember everything. Unless you're a Time Lord of course. And even then…

HAMLET

Where be your gibes now? your gambols? your songs? your flashes of merriment that were wont to set the table on a roar? Not one now, to mock your own grinning? Quite chap-fall'n? Now get you to my lady's chamber, and tell her, let her paint an inch thick, to this favour she must come. Make her laugh at that. Prithee, Horatio, tell me one thing.

HORATIO

What's that, my lord?

HAMLET

Who ist this fool?

HORATIO

Marry, I know not.

CLOWN

Does he mean me?

HORATIO

I think he does.

HAMLET

Ay you, sirah. What is your business here among the quick and the dead.

Puts down the skull.

CLOWN

[Suddenly dark and grim of countenance.]
I'll tell you my business, Prince Hamlet. My business is that skull.

HAMLET
 E'en this?

CLOWN
 The same. Yorick? Pah! That's no more Yorick than I'm a
 monkey's uncle.

GRAVEDIGGER
 Pay no heed, my lord. The fellow is... distracted.

CLOWN
 That's what you'd like them to think, isn't it grave digger.
 Or should I say – Lord Grathnave of the Erstwhile Collision.
 Hmmm?

GRAVEDIGGER
 But – how could you...?

HAMLET
 What sayest thou?

CLOWN
 And this skull is the missing second skull of the Fendahl. It's
 been lost for centuries.

GRAVEDIGGER
 Give it to me!

CLOWN
 No, I don't think so. No one should have the power this skull
 can impart.

> *The GRAVEDIGGER tries to take the skull from
> the CLOWN, who trips him with his scarf. The
> GRAVEDIGGER falls into the grave he was digging.*

HAMLET
What madness is this?

CLOWN
Ah well, that as they say is the question.

Exit the CLOWN, with skull.

Do you not know I am a Time Lord?
When I think, I must speak.

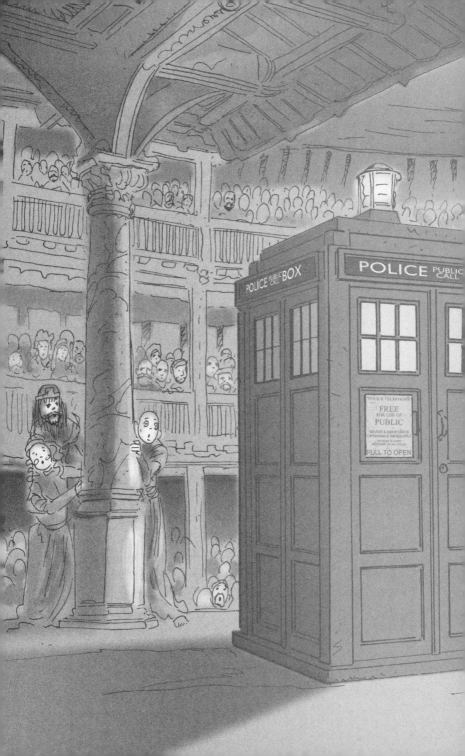

THE TIMELESS SHAKESPEARE

This section presents material generally taken from later years, right up to the present day, that would seem to elaborate on Shakespeare's apparent contention that the strange 'doctor' was somehow unbounded by the limitations of time. It includes post-Shakespearean material such as reviews of performances, critical reception, and academic analysis of the work.

TIMON OF ATHENS

Editors have often agreed that the text of Timon of Athens *we have is problematic. 'The copy from which the folio was printed had been mangled and interpolated' (Deighton, 1905). There has been much speculation as to whether this was caused by the theatre company or the printers of the 1623 Folio, but there is broad consensus that the play we have is incomplete.*

The first three acts in the Folio show us Timon, a retired soldier who has become fabulously wealthy. He is surrounded by flatterers who he makes extravagant gifts to, until his gold runs out. He appeals to his friends and the Athenian government for a loan, and is turned away by them all.

In Act IV we find Timon a starving beggar in the desert. Digging for roots, he finds more gold, but instead of rejoining society, he rails against his false friends and the Athenians, seeming to encourage the exiled General Alcibiades to attack the city.

In Act V the play ends suddenly, with Timon seemingly dead and Alcibiades rather abruptly the master of Athens. Several scenes appear to have been written by other hands ('There can be little doubt that the whole scene [V.III], which is quite irrelevant, is an interpolation.' Deighton, 1905).

For many centuries, Timon has been regarded as an incomplete or troubling masterpiece little suited to the stage. The Stationer's Register for 1618 contains notice of the intention by the printer Henry Dunwich to publish 'The Tragedie of Timon, or All For Gold'. The inclusion in the Shakespeare Notebooks of a partial and corrupt copy of this recently discovered and anonymous 'Bad Quarto' allows us to reconstruct some of the last two acts of the play.

ACT IV, SCENE III – WOODS AND CAVE, NEAR THE SEA-SHORE

Enter TIMON from the Cave.

TIMON

> Therefore, be abhorr'd
> All feasts, societies, and throngs of men!
> His semblance, yea, himself, Timon disdains:
> Destruction fang mankind! Earth, yield me roots!
> *[Digging]*
> What is here? Gold! Yellow, glittering, precious gold!
> Thy most operant poison. Ha! You gods, why this?
> Why, this
> Will lug your priests and servants from your sides.
> This yellow slave will knit and break religions.
> Come damned Earth, I will make thee
> Do thy right nature
> *[Covers the gold]*

Enter THE QUEEN OF THE AXONS

181

AXONIA
> Halt, good Timon. Why hids't thou thy glory?

TIMON
> Good Timon, say ye?

AXONIA
> Ay, good.

TIMON
> Good?
> There's no good in Timon.
> He is fly-blown, empty, spoil'd.

AXONIA
> But what of our gold? Does that not reflect good on thee?

TIMON
> Thy mirror'd glory is now a crack'd and
> Blackened stone. For far too late have I
> Learned the bitter lesson of thy gold.

AXONIA
> Did we not give thee all that thou desired?

TIMON
> Consumptions sow in hollow bones of men.
> A muddy soldier you found poor Timon
> And cleaned me quite in royal Danae's shower.
> That which you gave me I gave all away.
> I brought friendship, fame and power, but yet
> None held I sure. No thing on earth shall last
> Graves only be men's works and death their gain.

AXONIA
> I know thee well
> But in thy fortunes am unlearn'd and strange.

TIMON

I know thee too; and more than that I know thee
I not desire to know. Follow thy drum;
With man's blood paint the ground, rouge, rouge.
Axonia, I was by thy shining words beguiled
Yet let me rub the lustre from thy tongue.

Grabs a handful of dust. AXONIA retaliates

AXONIA

Hold you up your hands and your words, Timon,
They are dusty as thy dreams. We did but what you asked.
We gave you wealth, and you gave us Athens
We have gilded your city and there's naught but that
We touch, we own. To hold coin is but to be
By it possessed.

TIMON

Aye, I know that full well. A lesson learned too late.
But what now of Athens? Tell me of its fate.

AXONIA

Soon it shall be empty.

TIMON

'Tis empty now. Scoured of merit, of valour void.

AXONIA

We shall empty it of life.

TIMON

Oh, I like not that.

AXONIA

You like not much, or so I hear.

TIMON

> Hate all, curse all, show charity to none.
> And yet, but yet, even yet – shall Timon
> Let you unleash self-engendering death on all, Axon?
> A plague of gold that eats the souls it touches?

AXONIA

> Gold that will eat the world until the world's all gold
> It is for this that thou the world's soul sold.

TIMON

> Accursed queen, I'll stop thee yet.

Exeunt

ACT IV, SCENE IV – PLAIN OUTSIDE ATHENS

TIMON comes upon the General ALCIBIADES

ALCIBIADES

> What, Timon? I thought thou had washed thy hands of me.

TIMON

> No, gentle General, for I have more gold for thee.

ALCIBIADES

> Gold? Thou'rt rich again?

TIMON

> Aye, and so rich I know how poor I am.
> I shall give you Athens if you shall give me all its gold.

ALCIBIADES

> What?

TIMON

> The city's gold is cursed, well, the coin that came from me.

It is but fool's gold. A plague of gold full lethal to the touch.
See, see I sicken already? 'Tis death.
My coughing turns my coffers to coffin.

ALCIBIADES

I am sorry for you. What must I do?

TIMON

Beat at the city gates and demand my gold.
They'll resist you first, but soon, I tell you,
My gold shall bite. And you shall have both the
City and the gold. Beat down vile Athens
But touch not the gold.

ALCIBIADES

Do you so care?

TIMON

Not for Athens, nor for you. You are as honest
As any man can be.
If Alcibiades kill my Countrymen,
Let Alcibiades know this of Timon,
That Timon cares not. But if he sack fair Athens,
Then let him know, and tell him Timon speaks it,
In pity of our aged, and our youth,
I cannot choose but tell him that I care not,
But that gold, there's evil in it. Bring it back to me.

ACT IV, SCENE V – OUTSIDE THE GATES OF ATHENS

Trumpets sound. ALCIBIADES approaches with his Powers before Athens.

ALCIBIADES

Sound to this Coward and lascivious town,
Our terrible approach.

Sounds a Parly.

The SENATORS appear upon the walls.

SENATORS
General, what means you here? The city's sick
A plague affects the body politick.

ALCIBIADES
I bring a cure.

SENATORS
What?

ALCIBIADES
Give me your gold.

SENATORS
A costly cure indeed.

ALCIBIADES
The gold is poisoned. 'Tis death to all who touch it.

SENATORS
Explain.

ALCIBIADES
Damned senators and noble villains.
Know that I come as Alcibiades to claim thy
City and thy gold. You have done Timon wrong.
And, by his turn, he fears he has done you wrong.
He claims his gold is cursed, and would have it back.

SENATORS
If he wishes position again, why so did we offer it.
Why wants he back his gold?
We heard he does have more.

ALCIBIADES
You still want the gold and not the man.
I want neither man nor gold.
He pressed me the case the gold is cursed.
I surround your city. My force is great.
Give me Timon's gold.
And also give me those who have it touched.

SENATORS
We were not all unkinde, nor all deserve
The common stroke of war. These walls of ours
Were not erected by their hands, from whom
You have received your grief: Nor are they such,
That these great Towres, Trophees, & Schools shold fall
For private faults in them. Nor are they living
Who were ones that seized his gold
Shame hath broke their hearts.

ALCIBIADES
Shame?

SENATOR 1
Ay, shame.

SENATOR 2
And plague.

ALCIBIADES
Then there's my Glove, open your uncharged Ports,
Those Enemies of Timon must fall and no more.
Save the city and give me his damned gold.

SENATORS
Approach the Fold, and cull th' infected forth,
But kill not altogether.

The city gates open

ACT V, SCENE I – TIMON'S CAVE

Enter ALCIBIADES, with gold.

ALCIBIADES
  The Senators of Athens greet thee, Timon.

TIMON
  I thank them, and would send them backe the plague,
  Could I but catch it for them.

ALCIBIADES
  The city's mine, the gold is yours.
  Sour Timon, thanks. The plague has overcome them quite.

TIMON
  Touched thou the gold?

ALCIBIADES
  No. I saw them that held it sicken and so fall.
  Athens was but happy to give you all.

TIMON
  And do you give it all to me? Every last ingot?

ALCIBIADES
  All.

TIMON
  You are most human, and so must lie.

ALCIBIADES
  I kept back but a trinket. A mere souvenir.

TIMON
  Then that pocket token's mortal. Farewell,
  Alcibiades, none can save you now.

Enjoy your vengeance and your city,
But not long life. You have given that away
At a cheap price.

ALCIBIADES

Here, here, take your shining morsel.

TIMON

Too late. Away.

Exit ALCIBIADES

TIMON places all the gold in his tomb and lies atop it.

TIMON

Why I was writing of my Epitaph,
It will be seene to morrow. My long sickness
Of Health, and Liuing, now begins to mend,
And nothing brings me all things.

AXONIA

You have all our Axonite?

TIMON

Accursed queen, I have you overthrown.
My bed is gold, my sepulchre glows.
You heal me of this life. Reversed alchemy.
Transformed is Timon's gold to lead.
The city's saved for Alcibiades.
I wish them joy of each other.
Go, live still, Athens.
Be Alcibiades your plague; you his,
And last so long enough.

AXONIA

Bring out our Axonite.

TIMON

    I have thrown feasts with it.

    Now thy gold shall feast on me.

    A dinner of herbs, a lethal repast.

    Until we both are spent.

    Learn from me, Axonia. Man is deadly.

    Come not to me again, but say to Athens,

    Timon hath made his everlasting Mansion.

    What is amiss, Plague and Infection mend.

    Graves only be mens works, and Death their gain;

    Sunne, hide thy beams, Timon hath done his Raigne.

*TIMON lies down on the pile of gold and pulls the
mausoleum shut about him.*

Exit AXONIA

ACT V, SCENE II – ATHENS

Trumpets sound. Enter ALCIBIADES triumphant

Enter a MESSENGER.

MESSENGER

    My Noble General, Timon is dead,

    Entomb'd vpon the very hemme o'th' Sea,

    And on his Gravestone, this Insculpture.

*"Here lies a wretched Coarse, of wretched Soule bereft,
Seek not my name: A Plague consume you,wicked Caitifs left:
Heere lye I Timon, who alive, all living men did hate,
Passe by, and curse thy fill, but passe and stay not here thy gate."*

ALCIBIADES
  Dead Is Noble Timon, of whose Memorie
  Hereafter more. Bring me into your City,
  And I will use the Olive, with my Sword:
  We have learned that gold's a poison serpent
  And must poorly profit from our losses
  Make war breed peace; make peace stint war, make each
  Prescribe to other, as each others leach.
  Let our Drummes strike.

Exeunt.

Cry 'God for Rassilon!
Gallifrey and Time Lords!'

HAMLET'S SOLILOQUY

The earliest known copy of Hamlet's famous soliloquy from Act III Scene i in its final form is handwritten, but annotated with comments in the same handwriting. There has been some speculation that Shakespeare dictated the text, and the annotations were added by whoever took down his words. Earlier drafts of the text, with rather different wording, do exist – and indeed the annotations refer to these. But whatever the source, this is surely the earliest example of a critique of what is arguably the Bard's greatest work.

To be, or not to be[1] – that is the question:[2]
Whether 'tis nobler in the mind to suffer
The slings and arrows[3] of outrageous fortune
Or to take arms against a sea of troubles,[4]
And by opposing end them. To die – to sleep –
No more; and by a sleep to say we end
The heartache, and the thousand natural shocks
That flesh is heir[5] to. 'Tis a consummation
Devoutly to be wish'd.[6] To die – to sleep.
To sleep – perchance to dream: ay, there's the rub!
For in that sleep of death what dreams may come[7]
When we have shuffled off this mortal coil,[8]
Must give us pause. There's the respect

---

1 Well, I have to say this is rather better than your first attempt: 'To be or not, aye there's the rub…'
2 And what question is that, exactly?
3 I still think 'the songs and harrows' would be better. Still, it's your play. I suppose.
4 Mixed metaphor – I've warned you about these before, you know.
5 I keep reading this as 'hair' but that's my handwriting I suppose.
6 Now you're just cheating to make it fit the meter – worst example is "th' unworthy" later on.
7 If you just asked, I could tell you a thing or two about that, you know.
8 If you only knew how this phase will be used and abused. Even parrots shuffle off, apparently.

That makes calamity of so long life.[9]
For who would bear the whips and scorns of time,
Th' oppressor's wrong, the proud man's contumely,[10]
The pangs of despis'd love, the law's delay,
The insolence of office, and the spurns
That patient merit of th' unworthy takes,
When he himself might his quietus make
With a bare bodkin? Who would these fardels[11] bear,
To grunt and sweat under a weary life,
But that the dread of something after death
The undiscover'd country,[12] from whose bourn
No traveller returns – puzzles the will,[13]
And makes us rather bear those ills we have
Than fly to others that we know not of?[14]
Thus conscience does make cowards of us all,
And thus[15] the native hue of resolution
Is sicklied o'er with the pale cast of thought,[16]
And enterprises of great pith and moment
With this regard their currents turn awry
And lose the name of action…[17]

---

9 Long life? You don't know what you're talking about. I mean, you'll be dead before you're 60. Sorry, probably a bit tactless to mention that.

10 Do you honestly think anyone is going to understand this? Do you??

11 They may grunt and sweat a bit, as you say, but the Fardels of Astrogothicus Minor are actually very friendly and refined. Great chefs, the Fardels. They're especially good at open sandwiches, if that isn't a contradiction in terms (which it is).

12 To coin a phrase. Several phrases in fact.

13 It might puzzle you, Will, but it seems pretty straightforward to me.

14 That's just what I keep saying – but people will fly to me with their problems at the drop of a metaphorical hat. And scarf.

15 Thus… thus… thus – maybe vary it a bit??

16 I can see them in the back row, glazing over and muttering: "Oh look, he's off again." I mean, what is this all about? Tell me that.

17 You know, actually, that speech isn't at all bad. With a bit of work it could be really good. You've improved, you know. Entirely due to my influence of course, but you're making progress. I'm so happy for you.

ACADEMIC NOTES

As scholars are aware, the ambitious 'Shakespeare Project' set out to provide unrivalled academic notes and insight into each and every extant text by the great playwright. However, the project was abandoned under mysterious circumstances. The only work to have been completed seems to be this draft of the notes for a scene of Julius Caesar.

JULIUS CAESAR ACT II, SCENE II

CALPURNIA[1]

    Caesar,[2] I never stood on ceremonies,[3]

---

1 **Calpurnia** – Here Caesar's wife begs her husband to listen to her worries and supernatural portents that something bad is about to happen. See also similar forebodings in *2HVI* and *Macb*. ~~For instance, on my way to the library today, I saw Mr Silhouette again. If I were superstitious, I would assume grim foreboding~~. Whereas Q3 gives "Cal", Q2 gives "Calpurnia" her full title.

2 **Caesar** – Otherwise useless, the 1612 Duluth Quarto gives many additions to this speech, including the curious aberrant spelling of "Kaiser", which, if it really was derived from Duluth's own promptbook suggests a hard "C" pronunciation would be correct. For more on Duluth (1570–1612) see Appendix 6. Where I have included material from Duluth it is explained in the notes. ~~But I've left out the nuisance I had getting the thing out of the libaray. Yet here it is. In front of me.~~

3 **Ceremonies** – an incompetent line due to this polysyllable resulting in 11 syllables here. Gansard (1889) temptingly corrects this to "ceremon" offering a considerable improvement in internal rhyme with "stood on". However, it is an emendation I am reluctant to embrace owing to it being ~~nonsense~~ an uncommon usage.

> *The mind probe's the thing, wherein*
> *I'll catch the conscience of the king.*

Yet now they fright[4] me. There is one within,[5]
Besides the things[6] that we have heard and seen,[7]

---

4   **Fright** – Fear and forebodings are key concepts throughout *JC*. Fear is not seen as a weakness, but a proper reaction to a supernatural manifestation throughout, from Brutus down. The exception to this rule being Caesar, who refuses to state his fear of the future as being a reason to avoid attending the forum. ~~For example, Mr Silhouette is here again, sitting three rows over from me in the library. Yet again I can't see his face. But I sometimes think he's watching me. If I were a character in *JC* I'd take this as an ill omen and run down to the cafe for a bun. But no, I'm going to finish this scene today if it kills me.~~

5   **There is one within** – i.e. there is someone inside ~~with all the gossip~~ who can better inform you of these events. "Within" here both refers to inside Caesar's household and also indicates the area backstage – thus a clever staging pun. Interior and exterior domestic space in Shakespeare is most often seen in ~~you know I bet they've sold out of the marmalade buns. I really do like those. £3.50 a bun, though, which is possibly the greatest crime to happen in a library since the burning of the Library of Alexandria. And you don't even get a cup of coffee for that. Mr Silhouette is still there. Day after day. What is he doing here, I wonder? I should probably report him, but you really can't report someone for just staring into space in a library. That would be a literal thought crime.~~

6   **Things** – while it may seem a weak verbal choice to a modern ear, the word had a meaning closer to ghoul cf. "Back, thou hell-back'd thing!" (Simmond's *A Hey For Haringey*, 1586), or "I do consort with things spawn-born" (Landale's *Shilling For A Groat*, 1592). But, let's face it, really I hate this play. I'd have killed to get to do *Antony and Cleo*, but oh no, Granmer nabbed that one for himself. So I get the one where everyone tells the man not to go out, he goes out and dies, and then everyone spends the rest of the play talking about it. I dunno. Talking of things, Mr Silhouette's gone. I didn't even see him get up. Still, that'll teach me to check Facebook so often. Granmer's uploaded pictures of the yachting holiday he's on. Paid for by A&C, no doubt. There are pictures of Mrs Granmer on the boat. The barge she sat on looks in danger of sinking. Her idea of infinite variety is three different floral gowns and a sunhat with a broken brim. Granmer's not yet noticed she's reading Tom Clancy in every shot.

7   **Heard and seen** - "seen and heard" was a more usual usage, even then. Perhaps, given the reliance of *Tit.And.* on cliché in oratory, Shakespeare was here keen to reverse this to show the candid nature of this scene. Calpurnia is, of course, referring to events that the husband and wife have previously encountered offstage, although the audience would have heard them due to the use of the thunder-box. The odd thing, about Mr Silhouette is that I've seen him but I've never heard him. As though his footsteps are completely silent.

Recounts most horrid sights seen[8] by the watch.
A lioness hath whelped[9] in the streets;[10]
And graves have yawn'd,[11] and yielded up their dead;
Simonides'[12] fair warning has not been heeded[13]
Those who worry least have most to fear[14]
The strange and shapeless ones[15] draw hungry near

---

8 **Seen** – the repetition of "seen" here, together with the confused scansion and dislocated sense have given many to speculate that either this section is based on an inaccurate report, or that the compositor became confused when laying up this section. Or it could just be rubbish. Just saying. Wonder what Mr Silhouette's face is like? I can't ever remember what he looks like. Wonder if he's hot? Maybe he likes buns?

9 **Whelped** – needs to be pronounced as two syllables in order to sustain the blank verse. Come on everyone, it's your duty to help Shakespeare out here. See "Deadly Dactyls" in Smith's *Syllables & Sonnets* (1932).

10 **In the streets** – here Calpurnia refers to the streets of Rome as though a bad occurrence there portends misfortune to the ruler of the Empire. In other words, if I was the chief librarian here and I kept seeing Mr Silhouette hanging around, then I'd assume something was rotten in the state of Library. As it is, he's probably just a tramp. Explaining the damp, dusty smell around here. Like a flooded charity shop.

11 **Graves have yawn'd** – A recurrence of the mouth metaphors that run throughout *JC*. This is the earliest example in the play of the dead being given a voice, literally a mouth. See Plutarch (Appendix I) for a fuller explanation of these events. I say that, naturally, but I know full well that about 3 people will ever bother to look. One day, I swear, I'll leave the appendix out and see if anyone notices (Please see Appendix 17 ha ha ha). Gosh, Mr S is back. What's he doing here? While he was away, I sneaked a look at his desk to see if I could tell from his books. But no books left on display. Just a blank sheet of paper. The Mystery Man. Oh, talk about a coincidence, I swear he just yawned. Strange I can tell he yawned but still haven't seen his face.

12 **Simonides** – mythical king, blah blah Kingy Wingy blah blah sticky end blah blah. Gawd, so hungry. Buns buns buns. Anyway, these three lines are unique to Duluth, but convincing.

13 **Heeded** – an extra syllable breaks this line. Also the exact nature of Simonides' warning is obscure. What warning?

14 **Fear** – needs to be pronounced as a polysyllable, a not uncommon pronunciation at the time. The exact purpose of this line is uncertain. Perhaps it is warning that Caesar, who is not worried, should be. The whole play is an exercise in the foolishness of ignoring all the warning signs (find a way of phrasing this better).

15 **Strange and shapeless ones** – obscure, but Frederick (1917) postulates ghosts. This is a stable reading, backed up by the later appearance of

Fierce fiery warriors[16] fought upon the clouds,
In ranks and squadrons and right form of war,[17]
Which drizzled[18] blood[19] upon the Capitol;[20]
The city shook as though eternal lost,[21]

Caesar's ghost, which "draws hungry near" to those doomed to die out of revenge. Talking of strange and shapeless ones, oh yes, Mr Silhouette has changed tables. I can almost see his face now, as he's moved a row closer. Can't make much out because the lights aren't working down his end of the reading room.

16 **Fierce fiery warriors** – here Shakespeare translates the dramatic storm into the War In Heaven between the Angels which culminated in Lucifer being thrown burning down to Hell. This is, of course, as nothing to the sound the dogs made barking outside my window last night. I actually woke up to see what was going on. Not a sign of a pooch, naturally, but I wasn't the only one looking – someone in the street had stopped to look. I hope they don't think it's anything to do with me. The last thing I need is the Neighbourhood Watch complaining to me about my non-existent dogs.

17 **Ranks... war** Calpurnia here uses a vocabulary of a fighting army, foreshadowing the armies that will later wage battle after Caesar's death. Ooh, friend request on Facebook. Is it Mrs Gansard? Hope not. Oh. No, I don't know him. He seems quite nice though. Good boho scarf and all. Standing in front of a blue door. Ah, skip it.

18 **Drizzled** – Inexplicably, Duluth offers "pizzled" here. Idiot! No wonder he got stabbed.

19 **Blood** – And here begins the metaphorical rain of blood which proceeds in intermittent showers throughout the text. I should really do a weather forecast. Another friend request. Young man in a white jacket in front of that same blue door. What is this? The *Notting Hill* Appreciation Society? Can there be such a thing? I look up from ignoring it. Mr Silhouette has moved closer, but oddly, I still can't make out his face. Perhaps he's wearing a veil?

20 **Capitol** – i.e. Rome, although the capitalisation here quite odd. They've just announced the canteen's closing for the night. Right that's it. Off for buns. It really is a fiddle, having to check the Duluth back in and then dash down two flights of stairs.

21 **Eternal lost** – yet another comparison between the fall of Lucifer and the collapse of Rome. Also, what a waste of time. Cafe still open but no buns left. Nice Pavel said the last one had just been bought by an author of children's books. I asked him what he meant by that. He shrugged. "Really Odd Clothes. All children's authors dress funny. I think it is all the cats." Anyway, best push on. War in heaven lost, Eternal City forever denied to the fallen angel, diddly dum etc etc. Quite what it has to do with Caesar is a little puzzling – he just gets stabbed and everyone has a squabble, unless the idea is that the city state itself has fallen. Anyway, that reminds me. These lines are only in the Duluth Quarto, so I'd better go and check it out again.

The dome[22] quite crack'd asunder. Time was lock'd;[23]
The noise[24] of battle hurtled[25] in the air,[26]

---

22 **The dome** – While the ancient city of Rome did not have a dome, the idea here is clearly that the atmosphere around a city was created by the thoughts of its inhabitants. Cf. "The climate's delicate, the air most sweet" of Sicilia in *The Winter's Tale* III.i. Now, here's a funny thing. Got back to my desk with the Duluth Quarto. And sitting on it was a bun. With a note: "Enjoy." Mr Silhouette? I looked around. Still sat a few rows away. Inscrutable. Was he flirting with me? Or stalking me? Or completely oblivious to me? I just don't know. Perhaps I'll report him to the librarian. Once I've finished eating the lovely bun.

23 **Time was lock'd** – again, the idea that by acting outside the perceived political norms, Rome has removed itself from the normal affairs of time and exposed itself to the chaos that exists without rules. The bun was delicious btw. Even if, strictly speaking, you're not allowed food in the library. Had to eat it stealthily, like that old man who brings in tins of corned beef. Urge to go over and thank Mr Silhouette.

24 **Noise** – An important recurring theme in Shakespeare. Cf. "The isle is full of noises" in *Tempest* II.2. Occasionally cross-associated with taste and smell. Oh hell. YOLO. I'll go and talk to Mr Silhouette.

25 **Battle hurtle** – note the impressive internal rhyme and assonance. That was odd. I don't think the bun was a come-on from Mr Silhouette. He just sat there. Didn't look up. Just sat there. I assumed he was looking at his laptop, but no laptop. Perhaps Buddhist? Oddly, up a little closer, he's not wearing a veil. He does have a face after all. Anyhoo, I'll just check my dating profile.

26 **Air** – see note 22. The oddest thing. A message on my profile. From a man wearing The Worst Coat Ever, clearly from Asda's Forever Single range. Amazed he hadn't teamed it with a Simpsons tie. Anyway, the message read "Did you get the bun?" Whoa. That's Russian Mafia level stalking.

> *Rassilon, the Master and the Doctor
> are of imagination all compact.*

Horses did neigh,[27] and dying men did groan,[28]
And ghosts[29] did shriek[30] and squeal[31] about the streets.

---

27 **Neigh** – note the pun here on "nay". Even the horses know that something is wrong and can't ignore the signs. Just about to block Mr Bad Coat when he sends another message. "Good. I trust I now have your attention. I have to ask you something about the Duluth Quarto. It is vitally important." Surely he could just join the library and check it out? Or look at one of the PDFs of it online? But no. "What can I tell you that Google can't?" I ask. "Splendid," he messages back. (Who says "splendid"?) "You can tell me lots of things." "Why should I? Who are you?" "I am usually known as the Doctor, but you may have heard of me as Doctor John Smith?" Nope. "Of Smith's *Syllables & Sonnets*?" See note 9. And OMG. "That book is 100 years old." "Nonsense. 80, if it's a day. Listen to me you're in terrible danger, young lady—" Young lady? I blocked him at that point. New Life Rule: Accept no more Weirdo Pastries. Now then, back to Duluth.

28 **Dying men did groan** – Another internal inconsistency. We've already seen the graves yawn forth their dead, and time become locked, therefore we are in a Rome Eternal where death cannot happen (even though, in a scene or two, a whole lot of death happens). It is at this point that the Librarian shuffles over to pass me a note. He's wearing a look that's a mixture of "This is Odd" and "So Beneath Me". The note's in a really old envelope. It's got my name on it and a date. Inside it reads: "You're a very rude young woman. Check Appendix 4 to Smith's *Syllables & Sonnets*. 3rd Edition." You know what? I do. The book's on the open stacks (the unimportant stuff always is) and clearly no one's ever made it as far as the appendices – or did someone once drunkenly tell me something about them being odd? – anyway. I read Appendix 4. And then I come and I sit back down and I am very scared.

29 **Ghosts** – Surely that's nonsense? There are at least three reasons why that's nonsense. Only Mr Silhouette seems to have moved a desk closer. Which is, of course, not helping at the moment.

30 **Shriek** – A loud cry, but one portending disaster. Obviously, not a unique coinage. "It was the owl that shriek'd, the fatal bellman", *Macbeth* II.2. I've re-read that stupid Appendix and am now looking again at the Duluth Quarto. It's all obvious nonsense. I look around for the librarian, but he seems such a long way away. And the library is quite dark now. I guess it's getting late now. Home soon. Mr Silhouette is still here. Just us two. I find him quite comforting now that I can see his features. He looks quite friendly. And I can always call the police. Plus, still got WI-FI, so I guess I can always tweet the police. In an emergency. "Aaarg!"?

31 **Squeal about the streets** – This whole incident (from Plutarch) is quoted in "In the most high and palmy state of Rome /A little ere the mightiest Julius fell / The graves stood tenantless and the sheeted dead / Did squeak

All time has come,[32] to grab the prize, pay the price.[33]
O Caesar! these things are beyond all use,[34]
And I do fear[35] them.

---

and gibber in the Roman streets." *Hamlet* I.1. It had led some scholars to argue *JC* clearly predates *Ham*, and others that *Ham* clearly predates *JC* (see Appendix 2). An email pops up from Smith, John Dr. There's a little headshot of a completely different man. A lot younger. Is it some sort of society? A society of dons having a laugh? Trying to wrap me up in a bit of nonsense with some emails, a dodgy light and a bun? I glance at the preview of the email, "Please, you must get out of" and then delete it.

32 **All time has come** – a final weak addition from Duluth, and an 11-syllable line. It can be fixed by dropping the redundant and confusing "All" at the start. Should I have deleted that email?

33 **Prize... price** – the internal rhyme is perhaps the only merit of this addition from Duluth. I idly checked my Facebook. Just in case there are any messages. There aren't. I guess I'm a little disappointed about that.

34 **Use** – help.

35 **Fear** – worry. I've just noticed the wi-fi's gone. I look up to try and see the librarian. No sign of them. It's dark in here. I guess I'm almost the last one in. I don't normally mind that. And I'm not alone, after all. There's Mr Silhouette, sitting across from me now. Smiling. Funny that. I can actually see him clearly now. He's actually pretty handsome. I wonder if he's one of Mr Smith's Society? All part of the joke? Yes, that's it. His face looks familiar. I open the Duluth Quarto. Just to go to the frontispiece, the engraving of Duluth himself. It's that same face. I'm sure of it. I close the book. I'm going to look up in a bit. Just to check. When I dare. Because I know that he's standing over me. Waiting. And I'm terribly afraid. And just typing away. Because as soon as I stop

Hast thou forgot the alien Sycorax, who with blood and thunder was thrown into a duel?

SMITH'S SYLLABLES & SONNETS (Appendix 4)

Practical exercise: Rearrange the following into both blank verse and sonnet form:

Young Lady, you are in the most dreadful trouble.
The book that we spoke of I dare not touch.
If you leave its leaves then you can never leave.
It is printed on tragic paper. Paper that wefts and weaves
Around the reader's soul. It once was simply
Psychick paper, but it read too much of its
Owner's heart, and cream vellum blacken'd.
The writer is long dead, but enough of his dread
Soul is stored within those pages, hidden in lines
Waiting in words. Studying it is like
Turning on a tap to his black soul
Pouring it out into the world.
It was never meant to come into his hands
But now it is never meant to come into anyone else's.
Gentle reader, if this you see, my meaning now you know.

Once more unto the JARDIS dear friends,
once more; or close the time breach
with our Time Lord dead.

YE UNEARTHLY CHILDE

One of the stranger texts to have been discovered in recent years is what purports to be an account of a lost Shakespeare play titled 'Ye Unearthly Childe', although it seems unlikely that this was the title accorded the work by Shakespeare – if indeed it was written by him. The account is part of a set of diaries, badly damaged in the Great Fire of London (1666) and with no clue as to the identity of the author. The diaries were then lost until they were rediscovered in the 1960s. Even then, it was another fifty years before the possible importance of the text described was realised.

Sadly, the text is incomplete – there are just a few fragments with brief explanatory notes linking them. It seems likely that the sequences were transcribed during a performance, but unfortunately the earlier pages that probably explained where and when this performance took place were lost to the flames.

A ROOM OF LEARNING

Enter CHESTERTON and MISS WRIGHT

MISS WRIGHT
> She is a most unearthly child, methinks.

CHESTERTON
> Who, Susan Foreman? Aye, methinks 'tis so.
> That girl knows more than I will ever know
> Of Science: all I teach she knows.
> It is all child's play to her, no more.

MISS WRIGHT
> In hist'ry too, Ian, she knows so much
> Of ev'ry period, like she were there.
> I wonder how it is she knows so much
> And yet so little by the same degree.

CHESTERTON
>So little, Barbara? What knows she not?

MISS WRIGHT
>How many pennies in a shilling are
>Nor shillings in the pound.

CHESTERTON
>Can it be so?
>'Tis strange, I'm sure, but is not sinister.
>I know not whence your worry for her stems.

MISS WRIGHT
>'Tis not in that, though strange it is, I know
>My worry from another aspect grows.
>So far ahead of her classmates is she
>That I offered to tutor her at home.
>As soon as I suggested such a thing
>She grew afear'd, and said her Grandfather
>Would not approve of it.

CHESTERTON
>Well, he might not.

MISS WRIGHT
>But why grew she afear'd? It worried me
>So went I yesterday unto her home.

CHESTERTON
>You did what, Barbara?

MISS WRIGHT
>I found her home,
>The Secretary gave me the address,
>But when I got there, all that I did find
>Was just a junkyard, nowhere was a home.

CHESTERTON
> Perhaps you simply mistook the address.

MISS WRIGHT
> I checked it, Seventy-Six Totters Lane.
> No home is there, this junkyard all there is.

CHESTERTON
> Well this is strange and worrysome indeed.
> What purpose you to do about it now?

MISS WRIGHT
> She waits outside…

Here the extract ends. A brief note states that Chesterton and Miss Wright speak with the girl, Susan, and then follow her to the mysterious junkyard.

The following extract appears to take place within the junkyard:

MISS WRIGHT
> We saw her enter in, where has she gone?

CHESTERTON
> She can't have left; there's only one way in.

MISS WRIGHT
> But why should she come in here anyway?
> In this junkyard, which she, it seems, calls home?
> This muddied throne of junk, this rubbish isle,
> This mound of majesty, this seat of trash,
> This cluttered Eden, rodent's paradise,
> This fortress built by clutter for herself,
> Against inspection and the will to search,
> This realm, this I.M. Foreman's little world,

Discarded stone set in an urban sea
Which keeps within the circle of a wall
Which acts as moat against the scavengers,
This storage lot, this earth, this realm, this Junkyard?

CHESTERTON
Such clutter here there is would o'erfill
The wildest dreams of hoarding clutterers.
No purpose for this clutter can I see
Save as objects of curiosity.
See here, an ancient pram sits all alone.

MISS WRIGHT
The baby once within it now has grown.

CHESTERTON
Look here a lamp that once gave out great light.

MISS WRIGHT
And now, poor lamp, sit here through darkest night.

CHESTERTON
And here an organ, music came from this.

MISS WRIGHT
To hear music in this wasteland were bliss.

CHESTERTON
This vase held flowers once, and watched them bloom.

MISS WRIGHT
It now sits, barren, waiting for its doom.

CHESTERTON
This clock still ticks, its hands still keep good time.

MISS WRIGHT
Oh that its hands could Susan Foreman find.

CHESTERTON
    You're right; we must for Susan seek straight'way.
    No more distractions, no, no more delay.

MISS WRIGHT
    We cannot find her! Oh, where can she be?

CHESTERTON
    I know not. Wait! Do you see what I see?
    It's a Police Box. What's it doing here?

Again, it is unfortunate that the extract stops here – the pages badly burned – as the use of the term 'police' seems on the face of it somewhat anachronistic, as indeed do other terms such as 'junkyard'. Indeed, some academics claim the use of these anachronisms as sufficient to cast doubt on the validity and authenticity of the whole text.

    The transcription resumes with a short extract depicting the entrance of a strange Physician, who hears Chesterton knocking on the 'police box' as he searches for Susan:

PHYSICIAN
    A knocking? And my hearts are filled with fear.
    The Sound of Drums is sounding in my ear.
    The drumming calls my oldest, closest friend
    And with him come the deaths of countless souls,
    The massacre of millions, and much worse.
    His coming and the drumming are a curse,
    A shadowy conflux of evil kinds,
    And death will stalk to ev'ry place he finds.
    4 knocks.
    He knocks 4 times
    Nor slow'r, no faster.
    I pray it be some man, and not the Master.

CHESTERTON
> No sound within, and yet a sound without.
> Someone approaches, Barb'ra, hide yourself.

PHYSICIAN
> Who's there? Come out? Hmmm. I know that you're there.

The diarist's notes describe the next, missing section only briefly:

> '*Chesterton doth demand access to the Physician's strange cabinet, and when refused determines to fetch a constable of the watch.*'

There is clearly some missing explanation as the next extract then takes place within, as the diarist has it, 'a realm of mystery and dimension of wonder'.

MISS WRIGHT
> What majesty and wonder lies within?

PHYSICIAN
> Now quickly, Susan, back and shut the doors.

SUSAN
> Miss Wright? Is't you? And Mister Chesterton?
> What are you doing here?

PHYSICIAN
> You know them, child?

SUSAN
> They are my schoolteachers.

CHESTERTON
> What is this place?
> It's bigger on the inside than without,
> But that's impossible, it cannot be.

MISS WRIGHT

And yet with our own eyes this sight we see.
It's like a bush where roses bloom in snow
Or lightning hangs from trees like monstrous fruit.
It is hot ice and wondrous strange snow
This techno-Eden in a cupboard.
From outside all of this fits in four walls,
A tiny space, a cupboard in the street,
Yet here the walls are distant, shining white
With roundels on, and rising up so high
That they might stand above the highest trees.
This central console, on which controls gleam
Contains a central column grown from glass
And covered is with switches, lights and dials.
And if this gleaming cavern weren't enough,
I can see doors that would lead further in
Into the wilderness of this mad place,
The corridors of this strange powerhouse,
This techno-jungle, Eden of science.

PHYSICIAN

Yes all of that we see, well noticed, hmmm.
And further I would add to what you see
This is the dematerialiser,
Over yonder the horizontal hold,
And up there lies the scanner, these the doors
And that a chair with a panda upon't.
Sheer poetry, my dear, now do be quiet.

CHESTERTON

But it's impossible, I walked all round
And saw its size, or rather lack thereof.
It cannot fit within that small space.

PHYSICIAN

And yet it does, hmmm? Wizardry perhaps?

Following this, there are just two more short extracts, with no explanation for how either fits into the overall narrative. The first seems to depict the Physician – although he is evidently more like a magician – describing himself and his travels:

PHYSICIAN
    If you could the touch the alien sands, and hear
    The cry of strange, and wondrous monstrous birds
    That wheel through an undiscovered sky,
    Then would that satisfy you, Chesterfield?

CHESTERTON
    But say you speak true? Why would you be here?
    What drew you to this place and to this year?

PHYSICIAN
    Do you know what it's like to be exiles?
    Cast away from your home? Your life? Your world?
    My granddaughter and I are wanderers,
    Across the fourth dimension do we roam
    To whereso'er we please, except our home;
    That once discovered country, to who's bourn
    We one day shall return. Yes, one day shall!
    One day we once again will see those suns
    And walk between the silver leaféd trees
    And climb the snow-capped mountains of our home
    The Wild Endeavour continent to see.
    We'll talk with hermits of the daisy flow'r,
    Of time and tide and webs of history,
    We'll run with friends across the golden sands
    And dive into the shining crystal sea.
    We'll stand again upon that blessed plot,
    That sacred star set in the firmament
    As though it were a diamond in the sky,

The burning eye within Kasterborous,
(As unto you that constellation's known)
That well-kept garden of eternity,
That Ancient seat of Kings and Lords of Time,
That throne of Rassilon and Omega,
That Temp'ral jewel, that Gallifrey, that home
To which we shall return one fateful day.
Until that day we wander where we can
So question not what you can't understand!

The final extract, again offered with no explanation, would seem to close the play as the diary entry then ends, and the next concerns a very different occasion – an apparent account of Death itself visiting a village that fell victim to the Black Death.

MISS WRIGHT
    Has he gone mad?

SUSAN
    Not mad, but almost home
    For he, at heart, is a creature of time
    And loves to wander where so e'er he please,
    To learn the things no other being knows,
    To see the sights no other being sees,
    To live the days no other thing can live,
    To travel in the TARDIS every way,
    To take in all the gifts that she can give.

PHYSICIAN
    The dust on antique time would lie unswept
    'til whirling TARDIS time rotors it blow
    To let us see those sights as yet unseen,
    Where none can go before, we boldly go.

APPENDIX
THE LAST WILL

While it does not seem to fit into the general tone and structure of this volume, the following short story is included as an Appendix. It has no known provenance, but had been reproduced in various periodicals, journals and anthologies without explanation or attribution. Since it concerns both William Shakespeare and a mysterious 'doctor', it may indeed have some relevance to the preceding material.

> *I William Shackspeare of Stratford upon Avon in the countrie of Warr' gent in perfect health and memorie god by praysed doe make and Ordayne this my last will and testament...*

> *Item: I gyve unto my wief my second best bed with the furniture.*

'You're kidding.' Donna Noble accused the Doctor of kidding a dozen times a day. The thing is, he never was. Well, hardly ever.

'Nope.' The Doctor gave Donna a tiny push and she fell backwards. Onto the bed. Onto William Shakespeare's bed.

He'd been taking her on a tour of the TARDIS. Well, actually, she'd gone to grapple with the old-fashioned tea urn (emblazoned with a 'Votes for Women' sticker). After ten minutes she'd emerged without any tea to find the corridor around her had changed completely. She was utterly lost.

The Doctor had come for her. Eventually. He was wearing a different suit to the one she'd last seen him in. And he was holding a cup of tea.

She took the tea from him and asked about the straw in his hair.

'Oh, I've been having problems with a bed.' Sometimes the Doctor seemed to delight in coming up with sentences that had never troubled the English language before.

'A bed?' Donna's tactic on these occasions was to make a question out of the last two words.

'Yup.'

The Doctor had shown her the room. She wasn't sure what she'd expected the Doctor's bedroom to be like. But of course this was it. There was a child's wallpaper of rocket ships and little glow-in-the-dark stars were stuck to the ceiling. On top of a tottering pile of yellow *Beano*s was a stuffed toy owl. The most remarkable thing about the room was the bed.

'That's quite boxy.'

'Yes,' agreed the Doctor.

There was a lot to take in about the bed. For a start, it was the kind of bed you got in a posh mini-break hotel. It had the four posters. It had the curtains. It had the mountain of pillows and cushions. It had the elaborately embroidered sheets and the carved wooden sides. Somewhere, Donna surmised, was a honeymoon suite with a hole in it.

'Is this your bed?'

'Sort of.' The Doctor nodded. 'Well, I inherited it. It's actually William Shakespeare's bed.'

'Shakespeare's bed?'

The Doctor had nodded and suggested she try it. Then he'd pushed her onto it.

Donna Noble didn't bounce. She landed on the bed with a thump.

'It's got a straw mattress,' the Doctor explained. 'Don't worry – I've had it fumigated. No bugs. Zilch.'

Itching suddenly, Donna got off the bed. It wasn't just the imaginary fleas. She wasn't sure what was weirder – that William Shakespeare had slept on the bed, or that the Doctor did. She'd never really thought about him sleeping.

'Oh, I've never used it,' the Doctor said. He'd been looking at her curiously for a moment. As though waiting for something to happen. 'I'm just looking after it. For a friend.'

'The friend being Shakespeare? This being Shakespeare's bed?'

'Shakespeare's best bed,' corrected the Doctor. 'He left his wife the second best one.'

'Charming.' Donna rolled her eyes. 'I'm sure she was very grateful.'

'Well, she was actually.' The Doctor was being defensive. Clearly Shakespeare had been a good friend. Donna paused. Yep, there was another new sentence for the English language. 'The best bed went in the spare room. It was for guests, to show off.' The Doctor patted the elaborately carved sides of the bed. Nestling among the laurels were two masks, one smiling and one sad. Gambolling among the smiling mask were fauns and fairies, and a man with an ass's head. On a grassy mound a fat man sat laughing with a large woman holding a tankard and a platter of bread rolls. Clustered around the sad mask was a lonely young man holding a skull, a wild old man carrying a child, and a sour-faced king with a dagger. Standing on top of the masks were a boy and girl, reaching out to each other, but unable to touch. Yes, thought Donna, the whole bed was showing off.

Distracted by the carvings, she'd missed that the Doctor was busy talking. 'The second best bed, now that was Mr and Mrs Shakespeare's own. The one they slept in every night. Well, I say that, but Anne was a terrible duvet hog.'

'But I thought duvets weren't invented until—'

'Wedding present,' coughed the Doctor. 'So, William often sneaked off over the landing to sleep in the guest bed. Especially if he wanted to stay up late working on something. He'd sit up, scratching away and dreaming, looking out at the stars. Which is where the problem is. The problem of the bed. Why I have it.'

Donna's tea had gone cold. Mind you, if all of time and space were your car boot sale, of course the Doctor would have Shakespeare's bed in his room. What would she choose? Cleopatra's? She smiled at that, imagining some amazing gilded bed – not in her room in the TARDIS, but back at home, her mum boggling at it nestling in among the Ikea lamps and the Argos table. Donna would like that. Should she ever leave the Doctor. Just something to remind her every morning of how amazing her life had once been.

Lost in her reverie, she'd missed most of what the Doctor had been talking about. '… The problem was, you see, having sneaked in through his dreams in earlier life, they'd left a massive gaping psychic hole. It was fine when he shared a bed with Anne. Her dreams and worries (laundry lists, servant gossip, wondering what Will got up to in London) counteracted all of his. But when he slept in this bed, his brain never stopped churning out ideas, and that psychic hole in his dreams never ceased, pouring things out of his head…' The Doctor gestured at the bed, which suddenly looked ominous.

'You mean…?' Donna stared at the bed.

'Lethal queens, devious sprites, alien battle fleets—'

'Alien battle fleets?'

The Doctor waved a hand. 'Yeah well, one of us had had a bit too much sack that night. I'd just been thwarting the Cunning

Trees of The xlloxlttoxtl... and he'd just started Macbeth. Suddenly—'

'Oh!' said Donna. 'The bit where Burnham Wood walks to Dunsinane Castle?'

'Yup,' said the Doctor, patting the bed a little nervously. 'And as he slept, perchance to dream, a whole fleet of killer alien trees got stored in the bed.'

Donna stared at the bed. And now it seemed as though the bed stared back at her.

'All of his dreams are stored in that bed,' said the Doctor. 'So many potent thoughts. It's only because I fiddled with the mattress that the world wasn't overrun by fairies.'

'You fiddled with the mattress?'

'Well, you've heard of memory foam?' the Doctor said, pushing down on the mattress. The fibres inside rustled. 'This is memory hay. Kind of. Bulrushes of Lethe. Psychically absorbent and naturally hypoallergenic. Sweet dreams guaranteed.'

'There's a "but", isn't there?'

The Doctor nodded. 'Very absorbent. All of Shakespeare's dreams are stored in there. Problem is it's full. Crammed fuller than an iPod. And the next person to take forty winks on that...'

'And in that little sleep of death, what dreams may come?' Donna quoted.

The Doctor nodded. 'Which is why it's in my bedroom. After all, I never sleep in here—'

'I knew it!'

'I doze in armchairs.' The Doctor smiled. 'One life I'm going to wake up with terrible back pain.'

Donna ignored this. She'd spotted a problem. No Noble ignored a problem. 'But there's hay in your hair. And you said you'd been having trouble with the bed... and, back up the truck a minute, this is the most dangerous bed in existence, and you made me lie on it!'

The Doctor looked sheepish. 'Just testing,' he muttered.

'Testing what? That I didn't grow a donkey's head or go all Lady Macbeth?'

'Yurp,' said the Doctor eventually.

Donna gave him the stare she'd learned from Paddington Bear. It had made at least three different employers phone the temping agency to ask if someone else could come next week.

The Doctor jammed his hands in his jacket pocket, and then in his trouser pocket. 'I'd just been wandering by, looking for you. And spotted the door all open, and the bed all inviting, and there was a copy of the *Beano* and I was wondering how Minnie the Minx was getting on. I travelled with her once, you know. You see, Bash Street was being attacked by—'

Donna hated it when the Doctor fibbed. Or babbled. Now he was babbling and fibbing. 'You saw the bed, you fancied a nap and you nearly ended the universe?'

The Doctor blinked. For a moment she wondered if he was going to claim he and Dennis the Menace had fought the Cybermen. Instead, he said very quietly, 'Well, I don't think it would have been the end of the universe.'

'Course it would have been, you prawn.' Donna punched him softly on the shoulder. 'Cos you wouldn't have been around to sort it all out. While your alien trees and witches and nasty fairies would have been pouring out of the bed giving it all eye of newt, you'd have been asleep.'

'Ah.'

Donna punched him on the arm again.

'Stop that.'

'I never had an older brother,' said Donna. 'I'm making up for lost time. So come on, how did you save the world from Shakespeare's Bed?'

'Quite easy actually.' the Doctor took Donna's teacup from her and drained it. 'I turned the mattress over. Side B is empty.'

'And that's why you… you put me on the bed?'

'Yup.' The Doctor nodded. 'Testing.'

'And am I Queen Titania?'

The Doctor looked Donna up and down slowly. 'Not quite,' he said eventually.

'Pity.'

'Come on, fair Kate,' the Doctor said.

The two of them shut the door on the Doctor's bedroom and wandered away, leaving Shakespeare's bed behind them. And, as they walked up the winding corridors of the TARDIS, if either of them noticed the gentle patter of tiny feet and the ghostly, musical laughter that followed them, then neither of them said anything.

For God's sake let us sit upon the ground and tell sad stories of the fall of Arcadia.

The rest is... Silents!